RACE DECODED

RACE DECODED

The Genomic Fight for Social Justice

Catherine Bliss

Stanford University Press
Stanford, California

Stanford University Press
Stanford, California

Printed in the United States of America on acid-free, archival-quality paper

Library of Congress Cataloging-in-Publication Data

Bliss, Catherine, author.
  Race decoded : the genomic fight for social justice / Catherine Bliss.
    pages cm
  Includes bibliographical references and index.
  ISBN 978-0-8047-7407-9 (cloth : alk. paper) — ISBN 978-0-8047-7408-6 (pbk. : alk.
paper)
  1. Race. 2. Human genome. 3. Genomics—Social aspects. 4. Genomics—Moral and
ethical aspects. 5. Social justice. I. Title.
  GN269.B57 2012
  572.8'6—dc23

                              2011040727

Typeset by Bruce Lundquist in 10/14 Minion

For Mamoo

Contents

Acknowledgments

I AM GRATEFUL for the love, generosity, and guidance of many people. First, I thank the scientists who warmly shared their thoughts with me. I went into my first interviews with a dry list of practical questions and came out understanding the complex commitments and concerns at the heart of genomic science. I especially thank the people who showed me how they accomplish their research, and those who have continued to communicate with me as I have completed this book. The honesty and forthrightness with which these scientists approach the difficult topic of race gives me hope for our collective future.

Thanks are due to Orville Lee, who taught me how to think sociologically. If there is anything of value in this research, it is surely because he has shaped it. Sarah Daynes tirelessly met with me at my every request to hash out whatever piece of theory was troubling me. Oz Frankel continuously monitored my success and sparked a drive to write my own histories of the present. Whether we were talking about science studies or cultural sociology, Vera Zolberg's enlightening esprit has shined on me. Troy Duster exceeded my every expectation. From helping me set my course to introducing me to essential scholarship and its creators, Troy has been a pillar of my entire research process. His continuing support is a great inspiration to me.

During the early research process, I benefited from the critical attention of Ruha Benjamin, Manuel Vallée, Brian Folk, Gwen D'Arcangelis, and Nat Turner at the University of California, Berkeley, as well as the support of the

Department of Sociology at the New School for Social Research and of the National Science Foundation. At the New School, Ann Stoler introduced me to the power of ethnography at a critical juncture in my development as a scholar. Andrew Sproul guided me through the thicket of genomic science. Fanon Howell and Christine Emeran provided important intellectual support and relief from academic work.

I cannot begin to express my gratitude to my colleagues at Brown University. I would not be where I am without the friendship and stellar example that Sherine Hamdy has provided. Sherine was the person who motivated me to begin writing the book, and she was the last to set eyes on the manuscript, providing a close reading of it in full while in the field. Ipek Celik and Pablo Gomez blessed me with their supportive presence and careful reading of the manuscript. Bianca Dahl, Lundy Braun, Francoise Hamlin, and Alissa Cordner have brought joy to my life while contributing close readings of my writing. The Department of Africana Studies, Alpert Medical School, and the Science and Technology Studies program have provided an atmosphere of warmth and collegiality in which to work.

Phil Brown has been a most generous mentor. I don't know anyone with a finer combination of intellect, academic ingenuity, compassion, and deep concern for the people around him. Anne Fausto-Sterling has also been a wise guide. In addition to modeling how to lead a department with elegance and integrity, she has constantly reminded me why we write, teach, and take part in this most inspiring academic community. Corey Walker and Michael Steinberg have shown great concern for me from my very first months at Brown. I am grateful to be able to continue to grow with them.

The Cogut Center for the Humanities has been an important home to me, and I thank the 2009–11 fellows for all the clever brainstorming. The Pembroke Center for Research and Teaching on Women has also been an important foundation for me at Brown. Thanks go to Kay Warren and Elizabeth Weed and the 2009–10 fellows for conversations that have led me to see new research avenues.

My students have been equally integral to my success with this book. I could not imagine a better group of interlocutors. Richard Fadok deserves a special thanks for his efforts. The Andrew Mellon Foundation, Howard Hughes Medical Institute, and Brown University have provided generous support for my research and academic career.

Foremost, I am indebted to the members of my writing groups and to my informal mentors. Adrian Lopez Denis, Deborah Weinstein, and Deborah Levine met with me weekly for the better part of a year, and came through for all my deadlines, no matter how tight. I benefited immensely from their way of keeping the story alive. Ruha Benjamin, Aaron Panofsky, and Sara Shostak massaged my manuscript into its final form. They helped me to make its sociology sing. Ian Whitmarsh, Richard Tutton, Benjamin Hurlbut, and Andrew Fearnley read pieces of the manuscript at crucial moments. Daniel Lee Kleinman and Mathieu Albert provided limitless career advice. Ann Morning and Alondra Nelson showed the most beautiful kindness to me. I think of all of these scholars as my copilots, if not coauthors.

This book has benefited immensely from Kate Wahl and her editorial team at Stanford University Press. Their vision, diligence, encouragement, and eagle eyes helped make the book what it is. Thanks are also due to the anonymous reviewers of the manuscript for their exceptional comments.

Finally, I want to thank my family and friends for setting me on the right track from the earliest moment and providing joy and beauty throughout. My family both here and in Indonesia inspired me to love life and love to learn. I especially want to thank my mom for all the overtime and boundless care, and my dad, who has always been my intellectual base. Erika and Gramps are two individuals who have helped me live up to my name. I owe a special thanks to Papa, for loving me thoroughly through each and every second of my life. Who would have thought teaching me to read would lead to this! He and Mamoo are always in my thoughts and my heart.

All my love goes to my friends who packed the off-hours with immense enjoyment and meaning. They kept me balanced and sane through all of the long hours on this work. I especially thank Syama Meagher for inspiring me with her success and keeping me on course. I'm lucky to be able to grow with such a special person by my side. Kate Maxey's cheer equally fed my inner smile.

I met Luis Sampedro Diaz when I started this journey, and no one has transformed me more. I am happy to experience life's great adventure together. *¡Enhorabuena a nosotros dos!*

RACE DECODED

1   Introduction

What we've shown is the concept of race
has no scientific basis.
—J. Craig Venter, *International Herald Tribune*, 2000

Those who wish to draw precise racial boundaries
around certain groups will not be able to use science
as a legitimate justification.
—Francis S. Collins, *Cancer*, 2001

We could test once and for all whether
genetic race is a credible concept.
—Aravinda Chakarvarti, *Nature*, 2009

A GIANT FLATSCREEN with the words "Decoding the Book of Life: A Milestone for Humanity" blinked in the background. The velvety blue of the flag in the corner of the room took on nuanced textures as cameras flashed. On June 26, 2000, President Bill Clinton, flanked by genome mappers Craig Venter and Francis Collins, announced that the human genome had been mapped: "Today, we are learning the language in which God created life. . . . I believe one of the great truths to emerge from this triumphant expedition inside the human genome is that in genetic terms, all human beings, regardless of race, are more than 99.9 percent the same." Those present hailed genomics as the most transformative science in history—a milestone in human intellectual development, a sign of the arrival of geopolitical unity, and evidence of the essential fraternity of humanity. The most powerful scientists of the day joined Clinton in stating that scientific investigation into race would go no further. Genomics had once and for all closed the door on the idea of biological race.

When millennial headlines of the first map of the human genome declared the death of race in biology, no one suspected that by the end of the decade it would reemerge as the subject of intense genomic investigation. Speaking on prime-time television, across international news columns, in an array of public forums, and on Capitol Hill, the leaders of the Human Genome Project made the statement "there's no biological reality to race" a veritable national mantra. Pointing to humanity's minuscule 0.01 percent difference in our 3.2 billion nucleotides, scientists promised an end to centuries of scientific doubt, existential angst, and social struggle over racial difference.

It has come as a surprise, then, that since the mapping of the human genome, racial research has reemerged and proliferated to occupy scientific concerns to an extent unseen since early twentieth-century eugenics. President Clinton's celebratory remarks in 2000 certainly did not anticipate this outcome, much less that the renewed interest in racial research would come from within the inner halls of genomics itself. Human Genome Project reports of the summer of 2000 suggested that race was a dead issue in the sciences; yet, as early as May of the following year, newspapers were noting a new beginning for race-based medicine.[1] Biologists have since published more articles on race than ever.[2] In contrast to Clinton's seeming confidence that the debate about the biological legitimacy of race was over, a discursive explosion, along with a mushrooming of technologies developed in the service of testing, manipulating, or capitalizing on race, has made this decade of science one of the most race-obsessed ever. Scientists have scrambled to rewrite the book on race. Many have communicated a wide range of controversial views on race in major news media sources across the globe, views shared by powerful policymakers and public health organizations.

This book analyzes genomics' rapid shift from a science uninterested in race to one devoted to its understanding.[3] Examining the ways in which these scientific ideas are conceived, produced, and conveyed within the realm of science is crucial to comprehending shifting discourses and experiences of race in wider society. After all, authoritative sciences have bred humanity's most powerful racial ideas.[4] Furthermore, respected scientists have devised some of the most exploitative social policies based on their working understandings of race. Science and politics have long intersected to create tenacious systems of racial inequality—consider, for example, the role of zoology, anthropology, and ethnology in the slavery debates of nineteenth-century Europe and the United States; the linkages between evolutionary theory, Social Darwinism, and eugenics in

Progressive Era America; and the range of twentieth-century experiments that include Nazi twins studies and U.S. government-led syphilis and gonorrhea experiments in Guatemala and Tuskegee. The sociopolitical salience of scientific racial thought has been no less menacing in the case of newly emerging sciences than in authoritative ones. In fact, concepts of race have typically coevolved with new avenues in scientific innovation and expansion—looking back, we see each era's most vocal racial theorists at the helm of new scientific professional societies, editorial boards, state advisory councils, and policy leagues.[5]

Yet what is so fascinating about the case of racial science in the first decades of genomic research is that it arrives on the heels of three quarters of a century of policy designed to *prevent* research into biological differences in race. The United Nations Educational, Scientific, and Cultural Organization's (UNESCO) Statements on Race of the 1950s ushered in a series of collaborations between biological and social scientists who worked to dispel notions of innate racial behavior or inferiority. UNESCO and a host of other government agencies and professional associations followed these statements with declarations, meetings and seminars, and informational databases. In successive decades, powerful organizations such as the American Association of Physical Anthropologists, the International Union of Anthropological and Ethnological Sciences, the American Sociological Association, and the American Anthropological Association issued or updated their own statements on race, disavowing biological explanations of race and arguments for racial inequality. High-profile evolutionary biologists authored popular science books that abandon the notion of biological racial difference, and social advocacy groups used such statements to fight racism in their communities. The consensus at the end of the twentieth century was that to be properly antiracist one had to demarcate the social from the biological. Scientists who maintained a "colorblindness" or "race neutrality" suggested that by ignoring features and morphology like skin color when interacting and making decisions, scientific and otherwise, racism would abate.[6] This orthodoxy compelled scientists to look for alternate ways to represent human variation.[7] Explaining race became the domain of social scientific fields for over fifty years.

Despite this recent history, since the millennium's start we have seen genomics featured as the single authoritative source of racial expertise across a wide range of media. Headlines have run: "Race reemerges in a PC world," "Genome mappers navigate the tricky terrain of race," "Race is seen as real guide to track roots of disease," and "Race seen as crucial to medical research."[8] Months after the initial publishing of the human genome sequences, in June 2000, genomicists aired their views in internationally read periodicals like the *International Herald Tribune*, the *New York Times*, the *Financial Times*, the *Wall Street Journal*, and the *Economist*. Ancestry experts appeared repeatedly on NBC, ABC, CBS, PBS, and the BBC. A number of scientists espoused their personal takes on race in prominent biographies.[9] At the same time, unless quoted in relation to the genomic debate, purely social explanations of race all but receded from the news.[10]

Throughout the mass media genomics has come to be regarded as the new authority on race. A Google search on "science of race" and other similar terms in 2010 brought up thousands of websites where genomics was consistently touted as the corrective to the pseudoscience of past racial science.[11] Everyone from Wikipedia to the Health Department credits DNA with providing the ultimate truth about race. Influential science writers have rewritten human history in books dedicated to advancing genomics as the rubric for human variation and race.[12] A cluster of reality shows and documentaries with titles like *Motherland: A Genetic Journey* and *Who Do You Think You Are?* have sprung up to offer genomic solutions to ancestral lineages blurred by the legacy of slavery. This is no surprise in a world where the gene is a leading cultural icon.[13] Even leading scientists have published popular science books about race. Some have made films and toured the world delivering political messages based on their genomic findings. In science and medicine, but also the public sphere, genomics is seen as the leading expert authority on what it means to be human.

This dizzying change begs a few questions: Why has race once again become biologically important? How can an idea considered non grata in the biological sciences in 2000 become, by 2005, the focus of biology's pinnacle field? Why have the world's leading scientists embraced its study? What are genomicists saying about race, and how does it figure into practices in the lab, clinic, and beyond?

Inside Genomics

This book is about a field's struggle to craft an antiracist investigation into the biology of race. Genomicists responding to political debates over the ever fraught topic of human variation have formed a new scientific ethos and set of strategies to deal with the politically sensitive material with which they work. It was once commonly agreed among scientists that they should leave their personal histories outside their laboratory investigations. Now these same scientists are reflecting on their own understandings and life experiences to design studies that address racial health disparities, minority health, and biological processes associated with race. Many pragmatically and self-consciously use racial labels and even draw on their personal knowledge about group identity to recruit minorities. In doing so, they are building genomics as a comprehensive and ethically conscious new science of race.

This story highlights the convergence and the synergy between American science and politics during a time of rapid social change. In the early 1990s, a political doctrine of colorblindness gave way to the idea that differences should be celebrated, and that social "playing fields" should be opened to more kinds of political actors. Across the federal administration and public health, a paradigm shift occurred in which the leading strategies to battle racial ignorance and encourage diversity became minority inclusion and the acknowledgment of group identities and experiences. People working in institutionally distinct realms of science and politics have now come to unite over tactics like the strategic use of a biologically essentialist definition of race.[14] Many are reflecting on their own experiences to answer fundamental questions about race, formulating an antiracist activism from intimate life events. They are cooperatively interacting to create new research frameworks, expertise, and avenues for being human. The result is a widely accepted system of shared values and practices, and a consensus that race is meaningful socially *and* biologically.

Recent research has forecast the turn to what I call *race-positive*, or determinedly race-focused, genomics by analyzing the broader political framework of activism in which such research has emerged. Steven Epstein's examination of the inclusionary turn in American public health has shown that throughout the 1970s and 1980s social advocates, scientists, and government officials formed tacit coalitions to petition the government for the inclusion of women and minorities as subjects in biomedical research.[15] Their successes set in motion a cascade of policies to ensure that basic research and clinical trials were performed on a diverse array of bodies. These policies require scientists to perform categorical alignment between state classifications and research taxonomies. Recent research into legal and industrial norms has confirmed that such policies encourage race-based pharmacogenomics and diagnostics—for-profit endeavors that impact the way patient organizations and other advocacy groups manage the political terrain.[16]

Race Decoded follows the policy trail into genomic institutions, projects, and labs. I show that it is not just advocates and policymakers who are transforming biomedicine with a politics of identity; scientific elites have adopted this inclusionary paradigm as well. This book asks:

> How and why are scientists adopting racial classifications in their studies?
> What, aside from policy, motivates scientists to reconfigure their notions of race?
> What difference do understandings of race make for the science of genomics itself?
> How might genomic reconfigurations of racial difference change our social understandings of race?

Leading genomic scientists shuttle between popular notions of race, official racial standards, and data-driven categories of difference. In the lab, many adopt continent-based systems of ancestry or common lay racial categories to promote minority inclusion and make minority health a focus of research.[17] At the same time, these genome scientists also alter their research taxonomies to meet their immediate practical needs. Scientific research that integrates racial categories is not some mere aftereffect of policies handed down from Congress but is itself generative of new meaning around race.

This means that though scientists import policy-driven categories at the start of research design, they may also reflect critically on these categories and anticipate what social effects they might have. For the better part of a decade, social critics have been calling for genomicists to take greater responsibility for the social implications of their research.[18] Studies have emphasized how scientists uncritically draw on common lay notions of difference in their work.[19] Yet my conversations with an array of contemporary directors and lead investigators at the world's top genomic labs illuminate a *conscious* application of values at play in the changes in research strategies we see today. Elite scientists hold deep political commitments and impassioned views about race. Though their basic understandings of race differ, all support their beliefs with ethical and political justifications showing that they think through matters of race with social concerns in mind. Genomicists are using their knowledge of the political field to mount a fervent engagement with race and perform what they see as a civic duty.[20]

Like recent ethnographic research on genomicists, my findings show that elites personalize their participation in this new science of race.[21] Many scientists I spoke with discussed political reasons for going into human variation studies. They pointed to past racial experiences and ongoing antiracist activism in their home communities that have shaped their outlooks. Many also proclaimed a commitment to racial justice above all else, including scientific veracity and accuracy. Some called racial inquiry their "lifelong interest" or "personal passion." Scientists were quick to denounce the idea that science could be strictly objective and value-free. Instead, they intimated that science could be used for social activism. Stressing that their values shape the formulation of research interests and questions, a number of scientists attested to performing political acts even in their most basic scientific inquiries. Such an overt politicization of science allows scientists to cope with a politically fraught state of affairs. This shows a clear change from earlier scientists' ethos of a "culture of no culture."[22]

A politically conscious ethos in the production of scientific expertise has yet to be explored in the context of the new genomic sciences.[23]

With a gripping emotional gravity in his voice, asthma researcher Esteban Burchard recounted going from the barrios of San Francisco to a health disparities research division at Harvard University, where an asthma study turned up "a gene twice as common in African Americans [as in whites]." Burchard described this as "love at first sight," the moment when his lifelong commitment to health disparities, minority justice, and basic research decisively coalesced. Similarly, recalling the adolescent shock of moving from England to the segregated American Deep South, personal genomics specialist Joanna Mountain reflected, "I was interested and I was more concerned about the impact of racism *first*, before I was a scientist. But I enjoy science so much that I have come to value that world as well." These stories and many others show that scientists interpret their present work and respond to the present political terrain through lenses of ethical responsibility derived from consideration of their own racial backgrounds.[24] For them, race is both a negative symbol of legacies of injustice and a positive marker of community struggle and personal growth. Intimate knowledge of race serves as the basis for science activism—a mode of social action that rather than relying on protests or political campaigns, advances science as a solution for social change. This is not so surprising when one considers the force of minority inclusion ethics in the contemporary U.S. political landscape, of which genomics is now an integral part. Still a race-positive science was not anticipated by the planners of the major international genome projects of the 1990s. They had believed that avoiding the topic of race and the use of racial classifications would keep them sheltered from its political dimensions.[25]

Moving into the worlds of elite academic research centers, burgeoning federal health institutes, high-security technology innovation labs, and frenetic corporate headquarters, this book weaves analysis of genomic thought and practice across what experts are calling "the decade of the genome."[26] Unlike previous studies that have homed in on a specific technology domain, genome project, or lab, my study capitalizes on the field's tightly woven infrastructure and innovation stream to make a broad survey of its concepts and conventions since its emergence.[27] Following a "core-set" model of field analysis developed in the early Science Studies tradition—a methodology based on in-depth interviews with the cadre of scientists most influential in a particular scientific movement or field—I concentrate on the views and habits of the genomic professional elite.[28] Their narratives provide a window into the dominant values motivating the shift toward a genomics of race.

From April 2007 to June 2008 I interviewed thirty-six preeminent genomicists—the project founders, editors in chief, and professional society chair holders of the field. I also observed, shadowed, and interacted with many scientists in their labs, offices, classrooms, and conference rooms, and in an array of informal settings. Scientists were chosen for their leadership of international human genome projects and global epidemiological studies, their role in the invention and development of population genomics technologies used across the field, and their participation in field-defining public engagements on human variation like the publication of the human genome and the launch of direct-to-consumer genomics.[29] Almost all have led genomic research into plants, animals, fungi, bacteria, and viruses. Some have spearheaded the development of synthetic organisms. Most sit on the scientific and executive boards of pharmaceutical and biotech firms. Core-set analysis was specifically designed to get at black-boxed and unresolved knowledge,[30] the hidden and unsettled operational bases of science in the making, when it spans vast physical and ideological distances. In the case of genomics, there is a shortage of knowledge about the very scientists who make its hallmark decisions or the dynamics of strategy and decision-making that are behind this ostensible move toward researching race.

Because a consensus on race is still not complete, I also interviewed ten prominent critics and policymakers, three lab researchers, and two trainees, and I observed the creation, analysis, and interpretation of an ancestry estimation technology for three population studies. Finally, I examined the contours of debates with genomicists in an analysis of publications exploring the analytical validity of race for genetics and genomics.[31] This research allowed me to triangulate views and explore the cracks and fissures in the dominant account of how race should be scientifically addressed.

My research enabled me to see how actors from all camps have a vested interest in recuperating the term "race" to represent certain aspects of social standing. In this sense, genomics does not mark the reemergence of a prior science of race; rather, it is devoted to a new understanding of race—as a hybrid of molecular science, social epidemiology, public health, and bioethics. Within the field of genomics, scientists join social science experts in their efforts to recast race in historically conscious, yet politically empowering, terms.

Genomics has come to hold interdisciplinarity as a priority for the field. Openly valuing the subjective experiential rationales usually considered the mark of the humanities and social sciences, elite genomicists attempt to integrate a social science and bioethical posture into their basic methods. Scientists enlist social consultants for their projects and attempt to produce their own expertise on social matters. Social scientists and bioethicists also have instigated lengthy collaborations with scientists, research teams, and organizations. This process of mutual enrollment is an important factor in establishing the new science of race. Without it, entire projects fail and members of all camps lose social legitimacy, or "face," and the opportunity to give new meaning to the notions of difference and race.

Pragmatism, Values, and Norms of Science in a Biosocial World

Comprehending racial science today requires that we shift our framework for understanding the relationship between institutional mores, practical necessity, and personal values. Social studies of science have tended to elide analysis of the ways pragmatism and values coexist in its normative structures. When Max Weber explored the ethos of science in prewar America and Germany, he set a precedent for interpreting scientific commitment in terms of scientific objectivity. Though Weber discerned a scientific calling that cannot be reduced to the instrumental rationality he believed underpinned scientific work, he argued that the vocation of science attracted individuals motivated by a belief in progress and an enthusiasm for "self-clarification and knowledge of interrelated facts" beyond immediate or personal gratification. He also contextualized the rise of modern science in the West's transition from traditional society to legal-bureaucratic capitalism. Robert Merton developed Weber's idea by linking the modern scientific ethos to the same Protestant-based norms that precipitated a transition to a capitalist society. Merton later outlined a theory of norms considered to drive all science, and detailed a cultural reward system of scientific knowledge production based on *value-free* pragmatism. Investigating specific fields, others have continued to articulate normative structures that emphasize overt or tangible reward systems like citation in the literature and job promotion rather than ideological reward systems.[32]

In discussing genomics, Paul Rabinow has, in contrast, depicted a new, ethically endowed vocation of science, a form marked by

> a leitmotif among scientists, intellectuals, and sectors of the public turning on redeeming past moral errors and avoiding future ones; an awareness of an urgent need to focus on a vast zone of ambiguity and shading in judging actions and actors' conduct; a heightened sense of tension between this-worldly activities and (somehow) transcendent stakes and values; and a pressing need to define a mode of relationship to these issues.[33]

As Steven Shapin has put it, "What these people do, they do on a moral field."[34] Like Rabinow and Shapin, I find that optimism, charisma, dynamism, adaptability, and personal earnestness characterize the genomic ethos. However, in dealing with race, I would emphasize the ways collective responsibility to a specific set of racial values drives scientists and inflects their actions. In this sense, it is not some general belief in progress, some skeptical value, humanistic vigor, or vocational virtue that is at play. Rather, a highly contextualized set of norms and practices imbues this science with a commitment to correcting past injustice and establishing a new future.

The scientists I spoke with are open about the contingencies and limitations of their science, so open that they unrestrictedly discussed the *value-laden*, pragmatic nature of their inclusionary efforts in various projects. Many of the rationales they offered about minority inclusion and subject self-identification—rationales that emphasize respect for social communities and personal identities even if those self-understandings conflict with scientific data—are undeniably unscientific and threatening to the image of objectivity that the natural sciences enjoy. These scientists maintain that local support networks and community connections are of utmost importance to their work. They openly discuss details of their politically pragmatic sampling procedures without fear of being accused of playing politics, because they see genomics as inherently social and political and their role as values-based and ethically sound.

Such a variegated understanding of race has broad salience at a time when socializing around biological information is on the rise. Little more than a decade ago, Rabinow argued that genomics was creating a "biosocial" order. His forecast that "groups formed around [genomic classifications] will have medical specialists, laboratories, narratives, traditions and a heavy panoply of pastoral keepers to help them experience, share, intervene in, and 'understand' their fate" has clearly become reality.[35] Political interest groups and patient advocacy organizations dedicated to medical justice, social movements and community-based organizations petitioning for environmental justice, and government agencies interested in health welfare have turned to genomic knowledge to exact and administer resources.[36] The cottage industry of genome interpretation services has grown to be a thriving site of capital production.

Yet this exploration shows that scientists are not simply playing technoscientific handmaiden to a reordering of public ties. Drawing on their own experiences, memories, and racial values, scientists are themselves *biosocializing*. They are thinking through matters of race with their loved ones and themselves in mind and creating research agendas to promulgate specific values about race and science. These are actions social scientists have gone so far as to associate with the "ethopolitical" and "spiritual" nature of the contemporary moment, yet they have been all but ignored in the realm of scientific life.[37]

Thus far, most scholarly coverage of race in the biosocial era has left *scientific* biosociality out of the picture. Nikolas Rose, for example, has discussed racial biosociality in terms of a new somatic ethic driven by patient groups and the pharmaceutical industry.[38] Epstein has detailed the biosociality inherent in policymaking around public health racial enumeration. Jenny Reardon's study of Human Genome Diversity Project scientists, Michael Montoya's study of diabetes researchers, and Richard Tutton's study of British geneticists come closer to connecting scientists' racial values to their classification work.[39] All three have suggested that a colorblind ethic pervades these groups.[40] Only two studies have squarely placed the scientist in the biosocial context.[41] Duana Fullwiley has shown that pharmacogenomicist Esteban Burchard allows his a priori racial assumptions to guide his research into the population genomics of asthma. Alondra Nelson has described prostate cancer specialist Rick Kittles as evincing similar a priori assumptions as he markets his African Ancestry genealogy services to the public. While all this research suggests that biosocializing scientists are unaware of their assumptions, I find that genomicists are conscious of limitations in their present definitions and are actually motivated by that awareness to justify enlarging their public role.[42] The scientists I spoke with suggest that they are getting closer to a solution and thus require more investment in research that will shed light on the true nature of human variation.

I find genomic biosociality to be *reflexive*. Genomicists consider their role in setting the terms of societal biosociality and ask themselves what kind of biosocial future they want to produce. As one Jewish scientist, reflecting on the potential for genomics to verify Nazi assumptions, admits, "If I thought that it was in the best interests of people to fake my results, I'd be happy to fake it—I'd get myself in a great deal of trouble for saying that, but that's the reality." He and others in this study use their knowledge of the human genome to produce a specific set of biosocial relations. They consciously use racial classifications to recruit minorities, allow research subjects to self-identify, and attempt to create respectful descriptors with which to represent them. They manipulate racial classifications, despite knowing that there are strictly genomic ways to cluster data, because they value minority-appropriate strategies. The scientists here proactively publish on the merits and pitfalls of racial classification, spearhead research on hard-to-recruit minority populations, and engage in public dialogue on the limits of the race concept, all giving momentum to the case for *racial* biosociality. Though many of the scientists I met engage in such activity without having a clearly defined political agenda, all but one discussed ways to achieve racially sensitive goals.

Genomic elites head pharmaceutical and biotech companies and work on many privately funded studies; therefore, financial interests, careerism, and the like are also part of the field of their concerns.[43] What Dorothy Roberts aptly calls the "new horizon for profit" is ever present in elite deliberations and decision-making.[44] Yet the practical field in which these scientists work is inherently tied to its normative context. Scientists' preoccupations with their family and youth experiences show that research in a biosocial world involves open moralizing and personal enthusiasm about the work at hand.[45] Scientists have grown political and become politicized just as much as their research subject counterparts, and they are using their equipment to challenge the status quo.[46]

So while others have discussed life in a biosocial world in terms of new forms of sociality based on a public sphere immersed in bio-discourses, bio-products, and bio-expertise, I draw our attention to the way researchers themselves socialize around such developments.[47] My analysis goes beyond the actions of patients and research subjects to the creators of bio-knowledge itself. Scientists, who are themselves stakeholders of their research, form their subjectivity around the very knowledge they produce. It is not merely that new social relations form around new objects of research, but rather that the subjects of science—scientists themselves—recursively produce new subject positions around the objects of their own research.

Importantly, a new subject position and a new form of expertise has come into being: the figure of the genomic racial expert. The genomic racial expert is abreast of the latest in racial politics, has ties to minority communities, and launches large-scale studies on minority health. The genomic racial expert is the biosocial scientist par excellence, a scientist who reflexively considers the ethical implications of biological research from the first moment of inquiry. Unlike the scientists of yore, genomic racial experts don't just expose past racial science as bunk. They proactively seek funds for research that will benefit minorities and change the way society thinks about race. Albeit ambivalent this is a values-based source of expertise. Embodying such a position enables scientists to engage with the public in ethically salient ways that build social *and* material capitals that permit a redefinition of the field's reputation with regard to race.

Everywhere Inclusion

Perhaps surprisingly, the "unmarked" American—the white lab scientist—with presumably less stake in the redefinition of race, has been a major agitator on racial politics. Genomic racial experts are most often white scientists who, like minorities around the world, have been influenced by the history of race relations. Their growing concern shows us that race relations have had deep and lasting effects on people of all races, and not only those of the oppressed minorities. Race has been and continues to be important to the personal and professional development of scientists of all backgrounds.

Studies of race in America typically argue that race has created a bifurcated social landscape wherein racial minorities develop a heightened awareness of racial issues while whites have the privilege of ignoring racial inequality. W.E.B. Du Bois, for example, spoke of "double-consciousness"—an awareness by blacks that they are categorized as different and subordinate, and the pluralized viewpoint that results from simultaneously looking at oneself through the eyes of a dominant culture and through one's own experience.[48] Yet recent debates about the value of affirmative action and inclusionary measures have kept the educated public attuned to racial issues, even when they are members of privileged communities. Thus, more is to be understood about perception and consciousness in dominant demographics.

In order to perceive the nuances of race in America, and to understand its production in the world's leading centers of science, we need to extend our analysis of racial subjectivity to privileged and elite members of society. As scholars of race have argued, categories of difference that are formulated and experienced relationally must be studied as such.[49] The ranking categories of difference and the meanings attached to them apply to all members of a historical moment, and thus must be considered both within and across generations as well.[50] In this book, I view race as a belief system that produces consistencies in perception and practice at a particular social and historical moment. Scientists who have grown up amid a specific brand of racial activism, whether white supremacy or civil rights, and who work within similar policy climates share certain frameworks for thinking about race.[51]

Until the late twentieth century, racial difference was conceived as a fact of biology. Folk and expert notions of race both posited races as mutually exclusive biological populations worthy of different social and political statuses. Racial differences were ascertained by morphology and phenotype—visible structural features like eye color and hair texture. Race was conceived as typological, or capable of being characterized into discrete human types. Racial inequality was produced through a series of exclusions of all nonwhite types, including exclusion from first-class citizenship, labor markets, public resources, and entry to American shores.

Today, after a long battle in mainstream political and academic arenas, the idea that race is not a biological fact but a social structure predominates in intellectual circles. Minority inclusion is considered the salve to racial inequality. Scientists argue for racial inclusion in an idiom of optimism and empowerment that mirrors the broader culture of racial politics and Obama-esque campaigns of hope and social change. As social groups are recruited for comparative experimentation, difference is produced through *in*clusion.[52]

Though scientists challenge official strategies for creating inclusion, and many go rogue with their own taxonomies, the scientific adherence to the principle of inclusion creates a possessive investment in race. In a society that ordinarily silences racial discourse among the privileged, genomics provides a forum to reflect upon the impact of race and its personal traumatization among those perceived as "raceless." This is different from George Lipsitz's notion of whites' possessive investment in maintaining white racial difference.[53] In today's world of genomic science, there is a renewed focus on the value of race in a *multiracial* cohort of experts. While whiteness continues to hold a "cash" value that encourages whites to "remain true to an identity that provides them with resources, power, and opportunity,"[54] the white and nonwhite scientists described here struggle for the chance to redefine human taxonomy. Indeed, they believe genomics should hold the monopoly over ascertaining human categories. The white scientists in elite genomics laboratories are more invested in proliferating a positive sense of blackness than in protecting the biological semblance of whiteness. Furthermore, their redefinition of race involves drawing public attention to the social factors at play in the biology of race.

It is time to rethink the character, aims, and implications of scientific knowledge. Formerly, scientists used biological inquiry into race to naturalize social difference in essential biological difference. For example, lower social status, poverty, and lesser educational achievement, which resulted from racial inequity, were characterized as immutable and intrinsic properties of nonwhite races. Biology was used to obscure social explanations for race. By contrast, contemporary inquiry into race begins with the concept of social disparities and hierarchies and explores biological differences in order to correct those disparities. Instead of arguing that racial difference is impervious to social reform, scientists are expanding the definition of biology to include social factors; using their position to draw attention to inequalities; and applying scientific tools to create social change.

Postures that view racial science as inherently racist are thus untenable. Those who assume that all racial science is biological essentialism are missing the sociological nuances of today's science of race.[55] While studies of past science have fruitfully exposed the relationship between scientific taxonomies and social hierarchies—including their contribution to the production of racism in society's major institutions of law, education, the marketplace, and medicine—automatic dismissal of new developments fails to apprehend that scientists today are guided by the very norms and strategies that minority subjects in other realms have engendered.[56]

A better way of conceiving an enduring investment in race is in terms of "racialism."[57] Racialism refers to systems of racial beliefs that may or may not adhere to notions of hierarchy or biological essence. The term signals a move away from interpreting all investments in race as racist. It also helps us recognize the ever shifting ethical context for defining racial identities, a context that now produces genomics as a solution to prior racist racialism.

What I call *antiracist racialism*, or the idea that there is no rank to races but that there are nevertheless discrete populations worth studying, now prevails across science and society. Genomicists still enjoy the legitimacy of an objective "hard" science; however, they adopt antiracist measures that are widely supported by social scientists and the general public. Scientists and critics thus converge on strategies like allowing research subjects to self-identify their race and oversampling racial minorities to promote racial inclusion. They play out a "politics of recognition" with their science in ways that diverge from but also complement governmental acknowledgment of group diversity.[58] It is from a collaborative position that the field builds its reputation as the source of truth on human variation and is able to coherently produce a new science of race.

The singularity of today's racialism becomes clear on examination of prior antiracist politics in genetics.[59] In the postwar period, reputable evolutionary scientists such as Theodosius Dobzhanksy and Ernst Mayr suggested that experts in biology should use race only in the strict biological sense of populations at the subspecies level. Race was a matter of the invisible aspect of our biology: our genes. Objective genetic populations were to replace the typological, phenotypically characterized groups.[60] In the 1970s, a debate over race and intelligence came to a head between hereditarian scholars hailing from psychology and political science and geneticists studying human variation. At Stanford University, Nobel Laureate William Shockley used Arthur Jensen's theory about immutable differences in IQ to argue for a eugenic sterilization program for nonwhites having low IQs.[61] Luca Cavalli Sforza, Walter Bodwin, and their former student Marc Feldman, three of the world's most authoritative population geneticists, countered that there was no genetic basis to the behavioral differences at hand. Yet as with the geneticists before them, within genetics they promoted a strictly biological replacement for race: "Today, all continents of the world are inhabited by representatives of the three major human races: African, Caucasian and Oriental. The proportions of the three groups still differ considerably in the various countries, and the migrations are too recent for social barriers between racial groups to have disappeared."[62]

Though their work, and that of the population genetics mainstream, was devoted to elucidating racial divergence from the ancestral source, scientists were growing impatient with the connotative baggage that the term "race" carried.[63] Eventually, these scientists directed others to replace mentions of "race" with the term "population." This became a standard for the field.

By contrast, today's antiracist racialism is infused with an ethos of *political* justice. As later chapters will show, scientists use racial taxonomy to ensure equality in certain parts of the research and development process. Though most do not believe in a biological essentiality of race, they do strategically manipulate cultural beliefs in race. This is the principle of strategic essentialism that race and postcolonial scholars articulated as the temporary presentation of unified, essential groupness in order to gain resources for said groups.[64] Genomic essentialism is strategic in that those presenting the reified image of groupness know that there are vast differences within the group which are not supported by a biological notion of uniformity.

Analysis of the intersection of racialism and biosociality is sure to produce a sobering look at the way science and politics are coproduced.[65] Examining their relationship will show that race is not an epiphenomenon of science or technology but a special variable with a heightened constitutive force.[66] Contrary to the assumption that new disease classifications are displacing older systems of classification, racial taxonomies hold fast. Reflexive biosociality and antiracist racialism thus involve a more complex relationship between bio-subjects and bio-objects. Inasmuch as new technologies and expertise inform long-standing ideas about race, the demand for racial classification informs the development of new technologies, markets, and identities.

The questions most frequently asked thus far have been: Can a biology-driven redefinition of race solve society's racial problems? Or will genomics create an even more essence-bound version of race? These important normative questions will be duly addressed in the chapters that follow. I show that as genomicists produce alternative scientific perspectives on race, biotechnologies that can assist in the production of new DNA-based identities, and advice for policymakers, they fashion an activism that often neglects the core causes of racial injustice, such as institutional racism and structural inequality. Also, though scientists are responsive to criticisms of biological determinism, and though they adopt gene-environment models for understanding race, they nevertheless build genomics as a special expert science of race on the basis of its superior knowledge of biological ancestry.

However, this book seeks to understand something more. By treating science and politics symmetrically, I am interested in how the confluence of scientific and political conventions and norms becomes a multidimensional, multisituational *civic platform* for asking questions about human variation, designing research on difference, and creating meaningful change with science. In a time when the life of politics is increasingly trained on the politics of life, racial ideas and conventions are spanning wider distances and finding a home in unexpected places.[67]

My starting questions thus follow the trail of social processes: In order to fully understand how genomicists have adapted and resignified earlier categories of race, I first interrogate the history of these categories and how they have informed the emerging field of genomics (Chapter 1). I then trace the process whereby race became a norm for genomic research (Chapter 2) and show how public engagement also laid the foundation for genomics to become the new science of race (Chapter 3). I next ask, how do scientists conceive of race and what role do personal values play in producing the genomics of race (Chapter 4)? Further, beyond their personal convictions, what do these scientists actually do with race, and what does race do for them in the lab (Chapter 5)? Finally, in following scientists' public roles in genomics, I examine how genomicists position genomics as the ultimate expert field on race (Chapter 6). I conclude by returning to questions of racialism and biosociality in the context of emerging avenues of science and politics (Chapter 7).

Together these chapters reveal many counterintuitive angles into the lives of scientists, all the while showing that ideas about race are produced in the convergence of scientists' subjectivity and policy decisions, intimate values and market developments, ethical framings and technical practices, and disparate sites of antiracist activism. More broadly, these insights shed new light on the social coevolution of science and politics in a molecular age.

1 The New Science of Race

C ONCEPTIONS OF RACE are never a closed case of self-evident truths. "Race" is a cipher for a set of relational meanings—meanings of difference that are always underdetermined, pliant, and manifold. Although the term seems to connote a specific set of physical or cultural characteristics, there is actually no fixed referent for race. It is "constantly re-signified, made to mean something different in different cultures, in different historical formations, at different moments of time."[1]

Scientists make new meaning of race as they pragmatically respond to a range of official and unofficial policies, norms, and values, which are constantly being reworked. Knowing the historical background of policies surrounding official racial classification is but a first step in understanding how legacies frame present-day practice and how the political climate informs choices made in the lab and field.[2]

Technologies like haplotype mapping and principal components analysis—the state of the art in the production of genomic taxonomies—also shape how scientists define populations, compare their classifications with self-reported identities, and create new avenues for identity formation. The application of these technologies around the world in the present U.S.-dominated bio-economy has led to a hegemony of American racial taxonomies. Meanwhile, just as race-based medicine has hit the market, "whole genome sequencing" (technologies that map and interpret individual genomes in full) and the promise of a move to personalized medical healthcare are forecasting a de-parture from racial grouping. We must uncover what it means for scientists to create a biology of race that follows egalitarian and prodiversity goals, as opposed to colorblind or race-neutral goals, when that science is part of a larger economy of identity-based goods and services.

In the popular realm, genomics precipitates a greater emphasis on defining race with biological factors. New television and cinematic productions on race use genetic ancestry tests to fill in the blanks in histories lost to colonialism and slavery. They create an atmosphere where DNA technology is in public demand for solving identity quests and seeking political resources. These events signal a global interest in recreating social bonds based on new biological evidence.

These elements of the social milieu create a nurturing ground for race that has framed the past two decades of genomic science. They suggest that conditions are ripe for scientists to maintain race as an analytical lens in research, even as they attempt to offer new insight into human difference. Genomic efforts to understand race thread through the tangibles of genomic technology to produce a situation where investigating race and criticizing it from a biological standpoint go hand in hand.

The Backdrop of Official Classification

A long-standing system of governmental classification requires scientists to design all their research with official racial categories. These categories were established in 1977 at the behest of the U.S. Congress. In the wake of the Civil Rights Acts of 1964 and 1968, the Voting Rights Act of 1965, and a massive formation of new federal offices and agencies, Congress directed its Office of Management and Budget (OMB) to produce a government-wide set of race categories in order to monitor and redress social inequality. The OMB Directive No. 15 of 1977 directed federal agencies to use the U.S. Census' racial taxonomy to collect and monitor minority participation in public services.[3] The directive explicitly stated:

> These classifications should not be interpreted as being scientific or anthropological in nature, nor should they be viewed as determinants of eligibility for participation in any Federal program. They have been developed in response to needs expressed by both the executive branch and the Congress to provide for the collection and use of compatible, nonduplicated, exchangeable racial and ethnic data by Federal agencies.[4]

Despite the warning, committee reports show that Census race was still based on biological assumptions about essential, continentally clustered genetic differences.[5] Directive No. 15's initial classifications were: American Indian or Alaska Native, Asian or Pacific Islander, Black, and White. "Hispanic" was defined as an ethnicity.[6] Respondents have since been directed to report their Hispanic subgroup and mark more than one box if self-defined as a member of more than one category.[7]

Most significant about Directive No. 15 is the bureaucratic reach of this policy over American life. Census race was initially applied to collect population statistics on such varied topics as dairy product consumption, effects of prices, personal income, traffic deaths and rates, military deaths, minority business contracts, federal agency employment, voting age population, child abuse and neglect cases, drug law violation arrests, homicide rates, jail population, doctoral graduates, and employment and salaries of science and engineering graduates.[8] Today we can add to this list home mortgage disclosure, sterilization of persons in federally assisted family planning, homeless management, and more. The bureaucratic expansion of federal agencies and streamlining of information systems has exponentially expanded the directive's role. Presently, official classifications profoundly structure the work of agencies in terms of data collection and the administration of public services.[9]

Still, in public health it took nearly two decades to fully implement Directive No. 15.[10] Though Census race was immediately used in the reporting of vital and epidemiological statistics, not until minority and feminist advocates pressured the Department of Health and Human Services to take the inclusion of minorities in health-related research studies seriously was Census race incorporated into National Institutes of Health (NIH), Food and Drug Administration (FDA), and Center for Disease Control and Prevention (CDC) regimes.[11] Following a 1989 Surgeon General's report on the status of American health, Health and Human Services formulated the first set of top-priority national health goals that considered race. Data collection on racial health disparities—something notably absent from the first agenda—was designated as a key priority.

Drafted in the early 1990s, *Healthy People 2000*—the Health Department's decennial statement of national health objectives—set three broad goals: to "increase the span of healthy life for Americans," to "reduce health disparities among Americans," and to "achieve access to preventive services for all Americans."

> While significant improvements have been made in the Nation's health over the past decade, gains have not been universal. Therefore, many of the year 2000 objectives focus upon specific populations that have a higher risk of disease or disability compared to the total population. Minority populations are growing

faster than the population as a whole. Eliminating health disparities is of critical importance in the 1990s.[12]

Contributors from the Public Health Department, Institute of Medicine of the National Academy of Sciences, and over three hundred membership organizations and fifty-four state and local health departments stressed that an expansion of national health statistics on "non-black" and "non-white" groups was severely needed.[13] Inclusion of minority research subjects thus became a leading goal for all federal health agencies.

Amid Health and Human Services deliberations, in 1993, the NIH passed the *Revitalization Act*, a statute setting strict guidelines for the inclusion and surveillance of women and minorities in clinical research and clinical trials.[14] Building on previous policy statements that loosely encouraged tabulation by race, and on its recent establishment of an office for clinical research on women, the agency directed investigators to publish how their clinical research would affect women and minorities and to generate "outreach programs" to ensure inclusion. For the first time, cost was barred as a consideration over whom to include in a study. Minority exclusion was permitted only if the variables studied could be proven to have the same effect across national subpopulations. Until the *Revitalization Act* was signed into law, federal race policy had been vague and unenforceable.[15] Today the act continues to require all biomedical research—even research conducted *outside* of the United States on foreign subjects—to use Census race categories. NIH directors, advisory councils and staff, but also academic institutional review boards and peer review boards across the country have clear stipulations for enrollment and tabulation. The *Revitalization Act* has made federal race categories a prism for health research across the world and provided a population-defining context for subsequent research into genomic variation.

The FDA has followed a similar course since the mid-1990s. In 1998 it ruled that all new drug applications must "present effectiveness and safety data for important demographic subgroups, specifically gender, age, and racial subgroups."[16] This "demographic rule" requires race to be tabulated to ensure inclusion and reliability of data across the population because:

(1) Different subgroups of the population may respond differently to a specific drug product and (2) although the effort should be made to look for differences in effectiveness and adverse reactions among such subgroups that effort is not being made consistently.[17]

In other words, like the NIH *Revitalization* mandate, the FDA frames inclusion as *socially* relevant—important to consistent implementation of Congressional goals—and *biologically* relevant—important to the validity of findings across biological populations.

Since these major inclusionary policies were first established, each of these departments and agencies has revised and expanded its goals. In 1998 the Department of Health and Human Services formulated two in-house working groups on race: the Working Group on Racial and Ethnic Data and the Data Work Group of the Health and Human Services Initiative to Eliminate Racial and Ethnic Health Disparities in Health. In 1999 the department released a mandate stating that all department-sponsored research would thereafter use Census race classifications in "(1) data collection, (2) data analysis and interpretation, (3) data dissemination and use, and (4) data research and maintenance."[18] *Improving the Collection and Use of Racial and Ethnic Data in HHS* continues to guide all sponsored research to collect data on race.[19]

The NIH expanded its inclusion policy in 2000 and 2001. The upshot is an emphasis on "subpopulation" inclusion and surveillance for Phase III Clinical Trials.[20] The agency now requires investigators to "include a description of plans to conduct analyses to detect significant differences in intervention effect."[21] In essence, investigators must collect drug and treatment efficacy data using Census race. Even in the case of testing an intervention for which prior studies have proven no significant difference between subpopulations, the agency "strongly" encourages further data collection by race. This reorganization of trials into racial subpopulations affects more than the social composition of studies: *it creates racial sorting in the initial population design of trials.* Differential results by race are nearly a guaranteed outcome of such studies. Furthermore, contrary to OMB's original warning, biological conclusions are being drawn from socially defined racial classifications.

The 2001 revisions also provide stricter guidelines for implementation of Census race in circumstances where other categories might be applicable, such as foreign research. The NIH acknowledges the need "to explore collecting additional types of information on race and ethnicity that will provide additional insights into the relationships between race and ethnicity and health."[22] However, researchers are instructed to report such data separately from their main body of data and attach it to the standard enrollment form. The revisions stipulate that in the case of research on foreign populations all classifications are to be "designed in a way that they can be aggregated into the required [OMB] categories." In other words, alternate classifications may be reported, but they are not incorporated by the agency.

The FDA's policy revisions, made in 2003 and 2005, most explicitly exhibit the current drive to apply Census race biologically and globally:

> FDA recommends that the drug manufacturers use the OMB race and ethnicity categories during clinical trial data collection to ensure consistency in evaluating potential differences in drug response among racial and ethnic groups. Some differences in response to medical products have already been observed in distinct groups of the U.S. population. These differences may be attributable to intrinsic factors such as genetic differences; to extrinsic factors like diet, environmental exposure, socio-cultural issues, or to interactions between these factors.[23]

The agency now requires researchers to report racial differences in the *toxicological* effects of drugs and in disease susceptibility, and to offer Census race to research subjects in place of free-form self-identification. It maintains that improving patient safety and drug efficacy requires streamlined classifications.[24] The FDA's guidelines have enormous significance abroad, as the pharmaceutical industry increasingly locates drug development in developing nations.[25] Aware of the practical difficulty of imposing domestic classifications onto foreign populations, the agency offers a manual, *Reference Information Model Structural Vocabulary Tables*, to instruct investigators on allocating unruly responses. It "recommends using more detailed race and ethnicity categories when appropriate to the study or locale, but recommends that the OMB categories be identified for all clinical trial participants when submitting data to the agency."[26] As with the NIH Phase III Trial policy, the FDA explicitly excludes alternate data.

These policies mandate genomicists to consider human variation in light of Census race. Scientists are supposed to structure their studies by it from the inception of research design, sample by government classifications, and interpret and report data in a racial framework. Implementation of the government's taxonomy in genomics gives the federal system an imprimatur of science, while giving genomics an imprimatur of political truth. While the official policy is not the only stratagem scientists abide by, scientists must reconcile the government's policy with domestic and foreign conventions in their every effort at understanding human variation. It is a going concern for all scientists governed by these mandates to develop taxonomies that respect these most widely shared categories while finding new ways to conceive of human difference.

Technologies of Difference

Genomics and its signature brand of science activism attempt to provide bio-logical and social resources to the public in this fast-globalizing field of in-clusion. As molecular medicine moves from the "bench to the bedside," there are many technologies at play that empower scientists to study variation in a race-positive manner while also empowering everyday people to learn more about their ancestry and construct new racial identities. As such, objective and technical, but also subjective and personal, interests are embedded in the very technologies that establish genomics as the leading science of variation.

Until the 1990s scientists had three ways of understanding genes. They could study them indirectly through the observable properties of an organ-ism—phenotypic elements like skin color and height. They could hunt for genes individually by noting the prevalence of a gene in a disease-affected fam-ily and subsequently mapping its location on a chromosome. Or they could ex-amine the medical relevance of a known "candidate" gene through study of its position in the genome.[27] Accordingly, they applied emerging technologies to the same old questions that populated the earlier part of the twentieth century about the first divergences between the major populations of the world and population-specific diseases. Yet with the invention of the polymerase chain reaction, a technique that allows scientists to exponentially replicate a tiny supply of DNA, biology moved into what many call the "high-throughput," or rapidly automated, era of molecular science. Analyzing the human genome sequence with new bioinformatic computer programs led to an avalanche of data on cellular pathways, biological systems, and gene-environment systems. It enabled scientists to study common and chronic disease, such as heart dis-ease, cancer, diabetes, and stroke. It also allowed the mapping of other species, like model organisms used to study human disease, and microbial organisms responsible for the world's top killers—malaria, tuberculosis, HIV/AIDS, and meningitis. Within a brief time, genomics adopted an epidemiological method: the genome-wide association study. In genome-wide association studies, re-searchers compare the genomes of "cases" and "controls"—or diseased and healthy people. Scientists match cases and controls by race and ethnicity, so that only differences in disease status emerge from comparing their DNA.[28] Adoption of this method has brought the implementation of Census race into a more stark view, as scientists increasingly tackle common diseases that show deep disparities between racial groups.

These developments are occurring as "evidence-based medicine," or clinical science-based research, moves to "translational medicine"—medicine designed to bring benefits through the improvement of health management and patient care. Translational medicine combines social and biomedical science findings to produce better prevention programs and community-relevant health strategies. The turn to translational medicine encourages scientists of all backgrounds to take the social environment into greater consideration, including racial differences in socioeconomic status, work and living conditions, and health behaviors.

Genomics has turned also to the use of traditional population-defining technologies to help delineate ancestry in the interest of serving minority groups. Mitochondrial DNA and Y-chromosome technologies assess the nonrecombinant portions of the genome to map ancestral lineages and human migrations from the cradle of humanity and beyond.[29] The sequencing of genomic segments called "haplotypes" has occasioned scientists to group global populations into "haplogroups."[30] Private companies with names like "African Ancestry" and "DNA Tribes" sell genealogy kits that determine haplogroup membership, thus providing consumers a sense of belonging to an ancestral family. Different companies claim expertise about particular regions of the globe. Though mitochondrial DNA and Y-chromosome tests report only on the consumer's maternal or paternal lineage—less than 2 percent of an individual's ancestry—these kits have been popular among African Americans and others who have limited records of their genealogy. Recreational genomics has been lucrative, despite ample criticism that haplogroup membership cannot be accurately predicted without a comprehensive database of all the ancestral DNA in existence.

Admixture mapping and principal components analysis are two other technologies that capitalize on recombinant portions of the genome transmitted via haplotypes. Admixture mapping uses gene variants (alleles) that exhibit substantially different frequencies between populations from different continental regions to tease apart ancestral lineages and assign ancestral origins to factors of interest. Mark Shriver, a scientist residing at Pennsylvania State University, was the first to develop what scientists call "ancestry informative markers." Shriver's dissertation research, funded by the National Institute of Justice and later by grants from the NIH and the Keck Foundation, set out "to identify a set of genetic markers that would allow the confident determination of ethnicity, for use in a forensic setting."[31] Shriver originally planned to work with a set of ancestry informative markers that appeared over 50 percent more frequently in one ancestral group. But after scanning the genome for such "population-specific alleles," Shriver and others determined that markers exhibiting a roughly 30 percent difference are more common.[32] Scientists create an admixture profile from the combination of statistical probabilities that markers hold.[33]

Eventually, a group of scientists at the National Cancer Institute and the Broad Institute of Harvard and MIT discovered a way to infer ancestry at any chromosomal locus without relying on the informativeness of a few strong markers. Yet in the consumer realm, ancestry informative markers are still most commonly used. DNAPrint Genomics, a company that purchased Shriver's technology in 2002, has sold forensic and genealogy kits that present the admixture profile of a sample of DNA. Roots for Real, a European company that targets African American root seekers, launched a beta version of an admixture test in January 2011. Again, proportions are determined according to a prefigured set of continental groupings. DNAPrint has marketed itself as a company that helps customers change or confirm their racial identity. For example, it has targeted people who suspect they have indigenous American origins in order to sell them proof. Individuals can then attempt to register for tribal membership, petition for resources from tribal councils, or simply keep the knowledge to themselves.[34] Individuals looking to take advantage of an affirmative action policy or a legacy clause can also use their newfound ancestral profiles to support their college admissions. Though there are no published statistics on how many genetic genealogy consumers use tests to this end, in 2006 the *New York Times* ran an article exposing the practice.[35] This report stated that companies have encouraged customers to streamline their genetic information to Census race categories. Using test results for entry into a system of socially derived categories merits special attention, since this represents a direct translation of genomic data to lay signification. The fact that both companies and consumers find this practice legitimate demonstrates that the production of meaning about race is a collaborative effort between genomics and the lay public.

Principal components analysis software, programs that determine the most significant linear trends in multidimensional plots of genomic variants in order to group samples according to genetic ancestry, are more widely used in medicine. Some programs require researchers to determine how many groups they are interested in comparing, but they do not require any other identifying information about the populations they input. Most labs now employ some form of these technologies to establish the similarity between samples and pinpoint their ancestral origins. The widespread availability and routinization of these genomic technologies have made it possible to generate statistical populations in place of racial ones. However, slippages between what are considered "genetically related" versus merely "genetically similar" populations are rampant.[36] The result is often an overlap between traditional racial groupings and populations derived from genomic maps.

Though most of these technologies have been developed in the United States, genomic study of race does not affect just the United States. Indeed, concepts of difference are being restructured across the world as genomics enters new territory. Governments everywhere have made population genomics a cornerstone of health efforts and political campaigns. Nations such as Japan, Iceland, Finland, and Estonia have invested billions to create national population-based DNA biobanks that provide genomicists with unlimited supplies of domestic DNA. Mexico, India, and South Africa have established similar nationally protected genome initiatives to channel the nation's biological resources.[37] The United Kingdom and the United States have each launched a Cancer Genome Project dedicated to harnessing genomic insights from local cancer labs and centralizing data on open source websites.[38] The European Union has sponsored a 3D Genome Project to create intracellular DNA imaging and an international central database for genomic variation data. In the United States, the NIH Roadmap—a trans-institutional initiative of basic, translational, and clinical science—provides the funding to extend genomics to all agencies. Partnerships between Roadmap labs and other nations' labs have produced a series of international "-ome" projects like the Microbiome Project, the Variome Project, the Proteome Project, and the Epigenome Project.[39]

The changes that globalization brings are not simply about practices moving abroad. With the drop in price and size of machinery and reagents, local labs are revolutionizing practices and pushing for context-relevant applications. When technologies and therapies are made in new settings, practices themselves are transformed. Currently, many research avenues and technologies that are unsupported in the United States and United Kingdom are being pursued and produced elsewhere. China and Singapore are leading the way. Brazil and Russia are also building their economy around a cutting-edge biotech industry. This has direct implications for the hegemony of American racial models. On the one hand, we see an ensuing "me too" kind of antiracist racialism in emerging markets. In September 2010, China's Beijing Genomics Institute purchased 128 next-generation sequencing machines for a lab that focuses on diseases "specific to Asian populations."[40] The Genome Institute of Singapore's Martin Hibberd recently argued: "Most genomics research has been done on Caucasians based in Europe or the United States and we are only just starting to understand about how applicable these findings are to world-wide or Asian populations."[41] On the other hand, technologies in one region are being developed for multiple markets. As in the general case of patent applications, innovations and benefits must be translated from local terms. As Jonathan Kahn has shown, since 2008 the U.S. Patent Office has trended toward requiring patent applicants to include race in the section that legally defines the "metes and bounds" covered by the patent.[42] Thus research, development, and marketing are all conditioned by the U.S. standard.

This simultaneous thrust of inclusion and expansion has created an uneven playing field where Western politics and cultural investments in a continentally structured racial system inflect international genomic programs. As research comparing state spending on genomics has shown, the U.S. government funded the bulk of genomic research into the first half of the decade of the genome.[43] In fact, all of the international genome projects have been predominantly funded and directed by the NIH National Human Genome Research Institute. As Chapter 2 will show, while the Human Genome Project—the first genome project dedicated to creating a sample map of all the DNA in one human—did not use a racial sampling model but rather drew on already collected samples of convenience, the subsequent Human Genome Diversity Project used racial and ethnic categories to create a representative sample of the world's population. The Diversity Project was based on a vision of cataloguing the world's diversity through the study of genetically related populations. Though it was halted midway due to protests from minority communities who didn't want their DNA to be used for genomic purposes, the Genome Research Institute quickly launched a successor project, the Polymorphism Discovery Resource, to establish a DNA biobank structured by Census race. The International Hap-Map Project and the 1000 Genomes project, the most recent global projects, have created the most significant biobank yet. This database contains the DNA of thousands of individuals grouped by European, African, and Asian descent.[44]

In addition to the major sequencing projects, individual studies are overwhelmingly monitored from the United States. Adriana Petryna's multisited ethnography of clinical trials has shown that Western pharmaceutical outsourcing creates an uneven politics of recruitment where citizens of the Global South must present themselves as recognizable and amenable to Western frameworks.[45] Ian Whitmarsh's work in the West Indies has similarly found that U.S. inclusionary interests condition local West Indian healthcare regimes by calling for "black bodies" to serve as proxies for African American health.[46] The United Kingdom has followed a similar trajectory toward Census-based inclusion.[47] Indeed, many American policy shifts have analogues in the European Union multilevel governance system as well as the U.K. Department of Health. Consequently, as genomics promises governments around the world new tools for health governance, Western taxonomies continue to overshadow global health.

A Shift to Personalized Medicine?

Today the defining promise of genomics is a move to personalized medicine—a healthcare system based on the knowledge of each and every individual's personal genome. Pharmacogenomics, pharmacodiagnostics, and all kinds of genomic technologies have dropped in price so far that many experts assume they will soon be the basis for all of medicine. At Bristol-Myers Squibb, president of research and development Elliott Sigal said that in 2010 two-thirds of their drugs in the pipeline were influenced by genomics. Similarly executive vice president for research Marc Tessier-Lavigne reported one-third of drugs in the clinical trials stage and two-thirds of drugs in early development at Genentech as having been "enabled in a significant way" by genomics. Also, more researchers are using ancestry-defining technologies and conducting population-based studies than ever before. A 2010 review article in the *New York Times* described the latest genome sequencing machines as being the size of a computer printer.[48] The original sequencers for the Genome Project had to be housed in rooms as large as a football field. These newest machines cost a mere $50,000 and produce millions of gigabytes of data. It is now possible for smaller academic labs to own their own sequencers.

Indeed, as scientists have moved from the study of familial chromosomes to single nucleotide polymorphisms (SNPs, pronounced "snips")—variations in the individual A's, T's, G's, and C's that make up our hereditary code—genome sequencing has dropped so much in cost that an individual's entire genome can now be sequenced for a little over a thousand U.S. dollars. As I write this, a silicon chip with a digital interpretation of one's personal genome can be purchased on the open market for just under $50,000.[49] Scientists argue that once we have the complete sequences of each person's genome, race and all population-based medicine will become a vestige of the past. Yet today whole genomes continue to be interpreted vis-à-vis racialized reference databases that were established under earlier auspices of federal inclusion policy.

In fact, even though advances in the ability to sequence entire individual genomes might, theoretically, lead us to abandon racialized categories, these same advances have been presented in racial terms. When James D. Watson, Craig Venter, and anonymous individuals in China and Nigeria had their genomes sequenced, we heard about the sequencing of a "Caucasian," "Asian," and "African" being completed.[50] In November 2008, the Beijing Genome Institute published its first whole genome sequence as "The First Asian Diploid Genome."[51] In May 2008, researchers at Illumina reported they had sequenced the genomes of "three unidentified Nigerians," and in November 2008, they published the genome of a Yoruba individual from Ibadan, Nigeria, in the Hap-Map database.[52] In *Nature*, reporter Elizabeth Pennisi wrote: "As anticipated, the African genome had greater variation per kilobase than either the Chinese or sequenced Caucasian genomes, indicative of its ancestral status."[53] Since the appearance of these publications, the Genome Research Institute—in partnership with the U.K. Sanger Institute—and the Beijing Genomics Institute have embarked upon the 1000 Genomes Project, a project of whole genome sequencing in thousands of European, Asian, and African samples.[54] The project is using the HapMap samples to represent "broad geographic regions," so that unidentified Nigerians and the Yoruba come to stand in for "Bantu-speaking populations in Africa" writ large; Han Chinese turn into "populations in East Asia"; and samples from Utah become "populations in Europe."[55]

In the private sector, the direct-to-consumer personal genomics industry also suffers from its own racial slippages. When recreational genomics companies like Google-backed 23andMe turned to market genome-wide analysis in 2007, they carefully avoided making racial associations. Companies claimed to target the individual by tailoring genomic knowledge to the individual consumer and putting consumers' genomic data in their own hands. Yet personal genomics companies, all of which offer medical readings of the customer's DNA, make probabilistic claims about the individual's genome based on the litany of genome-wide association studies that often explain common disease risk in racial terms. Some companies ask clients to affiliate themselves with a preestablished racial or ethnic group so that they may focus their interpretations on a corner of their database. In these cases, clients can read their results only through a racial rubric. This raises the specter of race-based medicine, or drugs and technologies developed for specific races. The idea that each race has its own biological profile which merits uniquely tailored treatments mirrors the racial science of yesteryear, when scientists justified the maltreatment of racial groups and advised policymakers that enslaved and colonized groups were unfit for political freedom.

In today's climate, race-targeted medicine has indeed been brought to market. The first race-specific drug, BiDil, illustrates crucial contradictions. BiDil is a fixed-dose combination of an antihypertensive and a vasodilator, each of which is already available in generic form for patients with cardiovascular disease. When BiDil was tested for the treatment of heart disease in a multi-racial cohort, it was found to have significant effects only in people who self-identified as black. Later the FDA approved a placebo-controlled clinical trial to confirm BiDil's efficacy in blacks only. Though the drug is not tailored to an individual's genomic makeup, it has been heralded as a model for the personalized medicine industry. Supporters reason that the personalized medicine industry should use racial groups as target populations while the field improves its ability to address individual genomes, especially since the majority of epidemiological data on heart disease is reported by race. By this logic, groups from anywhere in the African diaspora can be treated as interchangeable until more is known.[56] Continental populations are sufficient "meantime" targets; thus Census race is a good enough approximation for genetic difference. Yet critics worry that personalized medicine will get frozen in its current racial guise.[57] Jonathan Kahn calls race "a useful tool for differentiating [pharmaceutical] products in a crowded marketplace."[58] In the case of BiDil, the drug's maker, Nitromed, held one expiring nonracial patent and one race-specific patent that would last thirteen more years. More recently, Forest Laboratories, the maker of a race-specific beta-blocker, was also approved for race-specific trials in African Americans and Mexican Americans, and subsequently filed for race-targeted patents. In the realm of diagnostics, AutoGenomics markets a minority-targeted genetic test for drug response variants related to Bristol-Myers Squibb's blood thinners Warfarin and Plavix. Pharmigene has also applied for a patent on a race-specific Warfarin test.[59] All of these companies have had financial incentive to position their drugs as race-specific.[60]

The New Politics of Race

The popular arena, like official policy and the biotechnological state of the art, also frames genomic science in ways that create a demand for biological solutions to questions of race and racial identity. While identity-based movements have typically sought to teach society about the messiness of social labels and their inadequacy to stand in for biological difference, a new political discourse has emerged in the media that links genomics and racial activists. Identity politics—or political activity based on the collective experiences and memories of injustice of identity-based social groups[61]—and genomics are getting a makeover in the process.

Since the civil rights, second-wave feminism, and gay and lesbian rights movements, the language of identity has dominated the public sphere. Identity politics has given marginalized members of society a voice with which to correct negative stereotypes and promote self-determination. It has evolved through health social movements fighting injustice in the realm of biomedicine.[62] Identity-based health social movements have traditionally struggled to publicize disparities in healthcare, medical treatment, and access to goods and services. They have also attempted to produce outreach and educational programs so that lay members of a community can formulate their own expertise about health and medicine.

In recent decades, identity politics has suffered a backlash. In the 1990s, political conservatives and policymakers like Newt Gingrich, Ward Connerly, and Dinesh D'Souza argued for a colorblind political system that would abolish race-based governance such as affirmative action. Soon after, liberal academics critiqued identity politics with the claim that political representations of identity had obscured differences and hierarchies within identity-based groups. But identity as a political framework for organizing has not retreated from the public. In fact, pride movements, group-specific tropes and genres in the media, and research centers studying and serving particular communities continue to proliferate. This precarious moment between colorblindism and identity-based sociality has produced novel solutions. As more and more diseases are perceived as genetic,[63] and as the public learns about genomics through its successes with diseases portrayed in the media as race-specific,[64] there is a growing demand for racial genomics to offer a way out of the inadequacies of identity politics.

Recent public broadcasts featuring prominent racial justice advocates are particularly telling. In 2006, Henry Louis Gates Jr., the African American studies scholar and director of Harvard University's W.E.B. Du Bois Institute, launched a two-part miniseries on the ancestral origins of high-profile African American celebrities. *African American Lives* attempted to showcase the history of the black experience; yet the series punctuated its episodes with stories of historical research in which DNA reports were framed as the ultimate word on ancestry.

Gates sought to create visibility for African American identity politics by placing African Americans at the center of the latest technological advances. He later enlisted Oprah Winfrey to further popularize African American genealogy in her own special feature, the culmination of which was the publication of the book *Finding Oprah's Roots, Finding Your Own*.[65] Winfrey's own interpretation shows how tests are being marketed as a salve to the legacy of slavery. She has said: "I feel empowered to say, 'This is who you are, this is where you've come from. You've come from strength and power and endurance and pain and suffering and triumph. You've come from all of that. And so imagine now how much more you can be.'" Winfrey and Gates have encouraged the public to think of DNA as an ultimate source of knowledge about self and past, and a way to heal from the violence of memory and its erasure.[66] The simultaneous denial and affirmation of race go hand in hand in these portrayals.[67] Tropes such as optimism and personal empowerment combine to form a powerful configuration of meaning.[68]

Gates has become the front man for an industry of recreational genomics.[69] From launching a genomic genealogy company of his own, to sitting on the board of a biotechnology firm, to being the first African American to have his whole genome sequenced, Gates has given genomics an imprimatur of political awareness.[70] In Gates's company's own words:

> [W]hen the paper trail ends and we have exhausted our sources, we are starting to look at something that our ancestors from Africa brought with them that not even the slave trade could take away: their distinctive strands of DNA. Because their DNA has been passed down to us—their direct descendants—it can serve as a key to unlocking our African past.[71]

These endorsements of genomics as a solution to racial quandaries foster an atmosphere in which the public is encouraged to see genomics as a social good. Making media appearances with notable racial advocates such as Gates has become a form of science activism that brings the field's image in line with those who have come out of identity politics debates as winners.

The first genetic genealogy companies on the market established such a pattern of promising to unlock customers' questions about their biological core, while providing the tools to recreate the public personae, social ties, and political narratives about which identity-based groups have raised consciousness. For instance, Oxford Ancestors—a genetic genealogy company established by Cambridge University's Brian Sykes—primarily focused on ancestral lineages of the British Isles.[72] Since its inception, Sykes has offered test results told through legends about founding mothers and fathers. Customers can read about their lineage alongside his best-selling book *The Seven Daughters of Eve*, *Adam's Curse*, or *Blood of the Isles*.[73] Social history is as much a framework for reports as the DNA data are. Other companies, such as Rick Kittles's African Ancestry,[74] have offered "Certificates of Ancestry," and now provide membership to online social networks.[75]

Increasingly, scientists have made their services more widely available by directly collaborating with filmmakers, artists, and community and public interest groups. Whereas the financial incentives are the basis for industry-based scientists—through niche marketing, companies ensure that a portion of the population will serve as a market base for products that ring in hundreds of billions of dollars each year[76]—many academically or governmentally based genomicists have also teamed up with race activists to put a racial justice spin on genomic ancestry reporting. The Broad Institute, for instance, intentionally stayed out of the production of *African American Lives* and the BBC film *The Motherland: A Genetic Journey*,[77] because, as members of the institute told me, they felt that genomic technology could not provide the answers people were seeking. They saw partnering companies as taking advantage of the public's ignorance about genetic variation. Yet as the Broad and other labs have advanced with whole genome sequencing, they have come on board with some of the same film production crews. George Church's lab at Harvard University has also gotten involved with mainstream genealogy by sequencing the entire genomes of Henry Louis Gates Jr. and his father. These labs have genotyped DNA for *Faces of America* and will provide sequencing services for a new prime-time television series on celebrity lineages. They too now speak in the language of racial benefits and slavery's salve.

The landscape for understanding race has been especially transformative for root seekers in the African diaspora, where consumers of genomic tests seek to align political and biological categories.[78] One example concerns Mark, a publicity agent in South London's music scene and one of three blacks selected to receive ancestry tests and a trip back to Africa in the documentary *Motherland*. Mark was initially disappointed to learn that a Cambridge University analysis of his Y chromosome indicated a Northern European genetic lineage, with no trace of African ancestry. He later heartened when a test of his matrilineal line, performed by African Ancestry, Inc., revealed a match with the Kanuri of Niger. As African Ancestry's Rick Kittles gave him the good news,

Mark's face lit up—a mix of eagerness to begin his journey and relief that his genome had the answers he sought. Kittles curiously studied Mark's face and said, "Just looking at you, you do have some features that are Kanuri." Filmmakers depicted how Mark's identity, formerly entrenched in a pan-African concept of race, changed. DNA purportedly illuminated a new, more essential way of constructing identity.

A year later, when the BBC producers of the series visited him, Mark was giving public talks on the importance of having a political basis for identity.[79] He explained, however, that DNA tests had been an important prompt for him to learn about the political genealogy of the category he now dons: that of "black Briton." Mark's televised experience suggests that genomics is transforming racial signification and representation in the public.[80] Scholars of race widely agree that racial groupness has heretofore been based on collective social experiences, historical legacies, and political action.[81] Yet as science is called upon to alleviate failures in historical understanding, who defines the group, and what defines groupness, is now uncertain.

Even in racial groups that have traditionally put a high value on kinship, DNA is replacing former notions of connectedness. Kimberly Tallbear reminds us that though many Native American polities require proof of familial descent from one or more registered grandparents,[82] or some proportion of maternal or paternal lineage, genetics is a new trope for evaluating kinship. Tallbear and Deborah Bolnick find that some polities are publicly encouraging individuals to use tests as proof of membership. For example, tribes that are unrecognized by the Bureau of Indian Affairs hope to inflate their numbers with individuals who can show that they have any kind of genetic similarity to tribal members.[83] When programs such as *Faces of America* recast Native American membership in terms of DNA, as that series did in episodes featuring celebrities Louise Erdrich and Eva Longoria, they help tip the balance toward a biological definition of race. Collective experience and action, but also kinship and filial relations, are subordinated to the unchanging essence of DNA.

Genomics to the Rescue

The field that today's scientists operate in is heavily marked by race. Whether actively making racial associations or carefully trying to avoid them, scientists must consider how they are going to characterize and interpret genomics with respect to race. Scientists work at the nexus of an array of institutional frameworks dealing with race. The scientists I chose for this study all hold multiple roles, including managing and leading labs and projects, counseling and participating in government forums, consulting to private firms, and sitting on the boards of pharmaceutical and biotech companies. Each of the institutions they work with is at cross-purposes: to continue with the dominant racial framework but to move beyond it as well. In this era of SNPs, microchips, and global partnerships, scientists continuously adapt to transfers of ideas, technologies, and labor across global regimes.

Today's Big Science is produced in the context of a quickening marketization. Though health policy committees require the vast majority of major projects and cohort studies to enlist the ethical views of an array of scholars, industry researchers, advocates, and even lay people in technology development and application, the public's uptake of new biotechnologies shapes the field in unpredictable ways.[84] In the case of genetic genealogy's popularity among members of the African diaspora who have little information about their ancestral past, the incentive to adopt genomic technologies is clear. However, these same technologies are behind a growing race-based forensic DNA industry.[85] Who is guiding, driving, conducting, and profiting from human variation research is a complex question that must further be explored.

To know where we are heading, we must understand the substrate on which scientists have worked out their orientation toward race and equality. We will now examine some of the field's major endeavors in the years that culminated in the Human Genome Project; this will illuminate the path that has made race-positive genomics second nature to science, medicine, and the general public.

2   Making Science Racial

WHILE IT WAS ONCE COMMONSENSICAL that the proper sensibility to counter racial discrimination was an emphasis on human universality and colorblindness, in the late twentieth century a shift occurred: the proper corrective to racial discrimination became racial *consciousness*, a disposition which requires the recognition of difference. As the field of genomics was born at the cusp of this major shift in larger social worldviews, it also became a field that most dramatically reflected it, and which granted this shift the aura of scientific legitimacy. From the late 1990s to the early 2000s, scientific elites steered genomics toward intensive investigation into race. They did so in responding to political pressures, seeking new funding opportunities, managing the field's image, and dealing with a burgeoning public interest in race. If paying too much attention to race had once been a mark of racism, now *ignoring* racial difference was heralded as a dangerous perpetuation of racial inequality. Genomic scientists adapted to meet novel demands for public accountability as notions of racism shifted under their feet.

This shift occurred within a broader civic move to include racial minorities and acknowledge minority status. As the political sphere became rife with race-positive measures, scientists moved from population-blind sampling—a protocol in which anybody's DNA was adequate material—to population-conscious, race-focused sampling. From a race-neutral to an inclusionary, prodiversity protocol, genomics has advanced its science in step with dominant American political positions.

The genomic turn toward race was not obvious from the beginning. The Human Genome Project of the late 1980s and early 1990s not only ignored the issue of diversity but also sidestepped the social redress of racism. In contrast, the turn-of-the-century Genome Project took up this banner as a central motivation and turned it into a template for other institutions and projects to follow. It also made community intervention a focus of its work. The recruitment of minority scientists and the development of culturally sensitive educational materials foreshadowed a commitment to, on the one hand, providing minorities access to genomic knowledge and, on the other, securing scientific access to DNA from racial minorities. In each of these cases, scientists made ethical choices about their sampling protocols as they reflected on the broader norms of social justice.

Examining the interplay of institutional policies and norms, I demonstrate how institutions with overlapping stakeholders and publics can converge on political agendas, creating a collective "institutional effervescence."[1] Many have noted the way that government, academia, and industry are now inextricably linked in their efforts to produce scientific knowledge.[2] Scholars describe this network as a "Triple Helix"—a multidimensional innovation system based on the wealth generation of industry, the novelty production of academia, and the public influence of government.[3] In the case of genomics, several key players have come together to create the civic platform that we see today. First, there are the NIH and the FDA, subsidiary agencies of the U.S. Department of Health and Human Services that oversee the recruitment of research subjects in biomedicine. As the primary sponsors of most genomic studies—even those conducted in the private sector to develop for-profit applications—their policies regulate the entire field. Some of their agencies have their own genomics labs and branches, which create influential methods and procedures; for example, the National Cancer Institute's Laboratory for Genomic Diversity, and the National Human Genome Research Institute's Ethical, Legal, and Social Issues unit (ELSI). The Laboratory for Genomic Diversity has been at the forefront of population genomic research and biostatistical development. ELSI has been the main sponsor of genomic ethical research from the field's inception in the late 1980s, and a forum for interdisciplinary dialogues. Second, there are biotechnology companies like Celera Corporation, Genaissance Pharmaceuticals, and Perlegen Sciences, which capitalize on investigation into the genomics of human variation. Celera, originally a division of the industry giant PerkinElmer, was the first company to sequence a free-living organism, and was the engineer of the "shotgun" sequencing strategy used to make the first map of the human genome.[4] Genaissance pioneered personalized medicine with its foundational technologies in haplotype sequencing. Perlegen, a spin-off of DNA microarrays producer Affymetrix, began as a gene chip developer for the pharmacogenomics market, but after taking a place at the cutting edge of SNP biotechnology, it joined the NIH in launching the International HapMap Project. Third, there are nonprofit institutes like the Broad Institute of Harvard and MIT, the Wellcome Trust, and the Centre d'Etude du Polymorphisme Humain. The Broad evolved from partnerships between the Whitehead Institute/MIT Center for Genome Research—a pillar of the Human Genome Project—and the Harvard Medical School-affiliated Institute of Chemistry and Cell Biology. With its $600 million endowment from Eli and Ethel Broad, it has become a world leader in basic genomic research. The Wellcome Trust is the United Kingdom's largest biomedical NGO, sponsoring over £600 million of research each year. The Centre d'Etude du Polymorphisme Humain (now known as Foundation Jean Dausset-CEPH, or just "CEPH") is a Paris-based genome center that houses DNA samples pertaining to two of the major international genome projects. It is sponsored, in large part, by the Republic of France.[5] Finally, there are academic research centers, such as the pharmacogenomics division of the University of California, San Francisco (what was formerly the Center for Pharmacogenomics), and the National Human Genome Center at Howard University, as well as individual labs housed in anthropology, biology, and genetics departments across the world.

Analyzing how these institutions collaborate—and how they share resources, strategies, and norms for recruiting, characterizing, classifying, and measuring human populations—shows that the emergent controversy over the biology of race consists of multidimensional partnerships in innovation and standard setting. Institutions and their members exist in a relational field, where the struggle for different forms of capital engenders structures of legitimacy for particular policy frames.[6] Scientific protocols come to reflect a litany of political values that are produced and institutionalized inside and outside the lab. Reverberations in policy and protocol seed new civic platforms that themselves feed back into assumptions about the meaning and utility of difference.

Mapping the Human Genome

The first international large-scale sequencing project was the Human Genome Project, founded in 1986.[7] The project was initially established by the U.S. Department of Energy to provide a reference tool to assess the damages of the atomic bomb to human DNA.[8] In 1988 the National Research Council and the Office of Technology Assessment elected the Department of Energy and the NIH to joint project management.[9] They recommended an annual budget of $200 million and provided Congress with an oversight plan for the project.[10] As Congress prepared to release the first fiscal year's funds, the Genome Project went global. In April 1988 an international consortium of leading geneticists from seventeen countries met in Montreux, Switzerland, to form the Human Genome Organization, which immediately established ethical, legal, and social issues as central foci of interest.[11] However, because the overwhelming majority of its members, and four of five primary sequencing centers, were located in the United States, the Genome Project developed its public intervention and sampling strategies under the auspices of the U.S. federal government.[12]

At the outset of the project, the U.S. government's health research inclusion campaign was in its inception. Colorblind ideology still dominated the public sphere, as exhibited by neoconservative challenges to affirmative action and rollbacks in social welfare. There were also countervailing attempts at racial consciousness-raising and minority justice. Behind closed doors an amalgamation of bureaucrats, scientists, politicians, and policy analysts were engaged in seminal debates over the demarcation of research populations and the meaning of diversity.[13] Social movements that were invested in the health of women and minorities, and a number of policymakers in public health and Congress, were just beginning to reap the spoils of a long battle to institutionalize research inclusion. Task forces on women and minority health were established, as were offices for special-populations health and preliminary inclusionary guidelines.

Still, although public agencies were required to use federal race categories in all offices and programs work, colorblind policies persisted across biomedicine and public health. There were two impediments to the implementation of these categories. First, the federal taxonomy was affixed with a warning that the categories were for *social* data collection purposes only. Second, a "one size fits all" paradigm, wherein research populations were assumed to be biomedically equal and one subject's biomaterial could stand in for another's, predominated across health research. Steven Epstein has argued that the habits of protecting vulnerable populations—groups with multigenerational histories of exposure to racism or sexism—and treating particular diseases with universal treatments created a climate in which twentieth-century researchers increasingly sought "white male bodies" from which to generalize.[14] Jenny Reardon has also shown that in the shadow of eugenics, geneticists were heavily invested in the belief that cleansing popular culture of the term "race" and transferring race quietly to the scientific domain was the best way for society to fight racism. That races were biological entities and equal in their value to science was a basic premise scientists did not debate.[15]

This paradigm was apparent in the Genome Project's initial sampling choices. The goal of the project was to assemble the sequence of a single set of human autosomes, plus an X chromosome and a Y chromosome.[16] This sequence would not represent all the human DNA in existence but rather a sample of the DNA that one human passes on to or receives from another. Project scientists needed an endlessly reproducible set of DNA to sample from, so they used cell lines already established by CEPH. These cell lines were created from people who reportedly originated in Utah, France, and Venezuela. In fact, the DNA in the database was known to possess unique mutations that conferred susceptibility to rare diseases; the assumption was that genomes were practically identical, even between diseased and healthy subjects.

During the Genome Project, Aravinda Chakravarti—2008 president of the American Society of Human Genetics, head of the Johns Hopkins School of Medicine's McKusick-Nathans Institute for Genetic Medicine, coeditor in chief of *Genome Research*, and chief advisor for the National Human Genome Research Institute—served as National Advisory Council member and chair of the subcommittee on the third Five-Year Genome Project plan. During my interview with him, he explained how they decided to approach human variation and whom to sample:

> I think immediately there was this argument saying, "Well, whose genome are we going to sequence?" Because everybody knows that these genomes are not the same.
>
> There were two views. One view was expressed by a group of people who said that we needed to sample different portions of the genome and different human beings to do it. And the other—which originally was a minority view but eventually became the dominant view, including the view that I hold and I think was the right view to have—is to say initially it doesn't matter. Let's not focus on the variation, because the point is not to study variation. You don't decide when a child is born and he is only a few hours old, as to where they are going to go to college. That's not what you think about.

Project planners knew that genomes possessed differences of interest, and that eventually they would study them, but for now the one-size-fits-all approach to health and medicine made deprioritizing the study of variation a viable option. Scientists felt they could postpone dealing with questions of difference until they had sharper tools for the job.

Yet scientists were, tellingly, anxious about how to manage the differences they would find. Chakravarti explained that the genetic maps being made at the time were "largely done on the European group," because it was "safe and because it was a practical matter of getting the map right." He and others said that project organizers stuck with the known reference populations because their ethical ramifications were known. As Chakravarti put it, "We wanted to sequence a human genome, not a *human's* genome." In a policy climate where all bodies were viewed as the same and inclusion was not an issue, allowing anybody's DNA to stand in for everybody's DNA made sense.

In the project's early years, genomics was struggling to become a field in its own right.[17] Protecting its autonomy meant shielding the field from racial dilemmas that had beset its parent fields of population and molecular genetics, genetic epidemiology, and evolutionary biology. As the field developed its technological capacity to deal with human variation, scientists began thinking about sequencing a broader spread of populations and sampling for variation. Later this would raise ethical quandaries about how to sample vulnerable populations—and how to go beyond the initial policy of, in Chakravarti's words, "not describing the sample at all, or very briefly, to having identifiable samples." Safety and practicality would be reenvisioned as a matter of enrolling the right minority groups and designing culturally sensitive recruitment programs. Providers of generous lines of funding would require researchers to study race closely.

The project's formal ethical dialogues and official goals also reflected a colorblind, one-size-fits-all logic. Early on, the Genome Project established its ELSI branch to address issues raised by new knowledge about human biology and the circulation of personal genetic information. ELSI's goals in the project's first Five Year Plan did not address issues of diversity and population representation, nor did they query how groups would be affected by genomic knowledge. They focused instead on potential harms to the individual, such as breaches in privacy and confidentiality, potential discrimination by insurers or employers, and managing personal genomic risks and susceptibilities.[18] In the project's second Five Year Plan, an even vaguer list of concerns was offered: "(i) Continue to identify and define issues and develop policy options to address them. (ii) Develop and disseminate policy options regarding genetic testing services with potential widespread use. (iii) Foster greater acceptance of human genetic variation. (iv) Enhance and expand public and professional education that is sensitive to sociocultural and psychological issues."[19] The last two goals provided a potential space for dialogue about the biological significance of race, and the impact the project would have on popular notions of race, but made no explicit mention of race or diversity issues. When the National Center for Human Genome Research—the precursor to the National Human Genome Research Institute—released its first retrospective, deputy director Elka Jordan declared the ELSI priorities of privacy, discrimination, and genetic testing dissemination as critical to the project. No word was raised on race.[20]

The race of the individuals who donated their DNA was thus not at issue in the earliest years of the Genome Project, and the idea of representing major populations was not discussed.[21] The project's protocol specified drawing samples from the preexisting database, anonymizing them, and using the most robust samples for as much of the sequencing as possible. Scientists conceded that they limited sampling to populations in the United States, where institutions were in place to ensure the procurement of informed consent. As a result, the privately and publicly derived sequences alike reflect largely the DNA of the two American men of northwestern European origin who happened to donate the most robust DNA. Genomics in this stage was, for all intents and purposes, race-free.

Racial Revitalization and the Making of Diversity

One way to interpret transinstitutional patterns in policy and protocol is through the lens of "biopolitical paradigms"—knowledge frameworks that reflect the intersection of governmental and biomedical policy interests.[22] This allows us to view such patterns from the angle of scientific conventions, research interests, and the political values that scientists enact in research practices. Yet it is also beneficial to take stock of the larger "cultural unconscious" around race; namely, the *racial* paradigm that obtains beyond the domains where official policies rule.[23] From this perspective, protocols are modeled not only after trends in policy but also after trends in racial activism. Because health equality has increasingly been fought for in a racial idiom, genomicists like other social actors have successively built their projects as "racial projects"—political projects that attempt to "reorganize and redistribute resources along racial lines."[24] In a newly charged political arena where inclusion and diversity reign, scientists have promptly relinquished population-blind sampling in favor of population-conscious sampling and overt deployments of political concepts of race.

As genomics embarked upon its no-labels, representation-neutral project at the end of the 1980s, a series of debates in Congress and the Department of Health and Human Services was turning the tide from the one-size-fits-all paradigm to a race-positive "inclusion and difference" paradigm—a framework for allowing inclusionary categories to lead the recruitment and measurement of diverse populations in research.[25] An ethos of multiculturalism was fast developing within the general public, and the underrepresentation of women and minorities was becoming a widespread political concern. Colorblindness was at once peaking and coming under attack in the broader culture. The cover of one issue of *Time* magazine presented a vision of the new face of America: a lightly tanned, race-free amalgam of the seven major ethnicities in the U.S. population.[26] Books with titles like *Postethnic America* pleaded for a new cosmopolitanism based on voluntary cultural affiliations.[27] Yet liberal intellectuals who argued that America had headed into a "postracial" condition were accused of ignoring the daily prejudice that nonwhites faced.[28] Many veteran civil rights supporters were criticized for whitewashing the institutionalized privileges afforded to white Americans through the selective issue of government aid, military grants, bank loans, and housing opportunities. In public health, social justice advocates and social epidemiologists argued that severe health disparities cut across the American populace. Political leaders agreed that the state would have to create an inclusionary health policy akin to affirmative action for the healthcare industry.

Health and Human Services' *Healthy People*, foreshadowed these changes. As noted in Chapter 1, the Healthy People program was originally devoid of any mention of race or racial health disparities. But in 1990 the department issued a plan determined to "increase the span of healthy life for Americans," "reduce health disparities among Americans," and "achieve access to preventive services for all Americans."[29] *Healthy People 2000* aimed to establish minority health statistics based on the quadrilateral Census taxonomy of American Indian or Alaska Native, Asian or Pacific Islander, Black, and White.

As talks proceeded on Capitol Hill, genomicists at a number of leading academic institutions began discussing the possibility of an auxiliary human genome project dedicated to human variation. In 1991 scientists announced plans for a Human Genome Diversity Project. The National Human Genome Research Center, the Department of Energy, the National Science Foundation, and the National Institute of General Medical Sciences funded three planning workshops for the project, and the National Human Genome Research Center and the Human Genome Organization voiced interest in committing future funds. Human Genome Organization president Walter Bodmer went so far as to call the Diversity Project "a cultural obligation of the [human] genome project."[30] The field was waking up to the idea that genomics would have to include minorities in its primary research programs, and the Diversity Project seemed like a good start.

The project was proposed by two of the most reputable scientists in the genomic parent fields of biochemistry, molecular anthropology, and population genetics: Allan Wilson at the University of California, Berkeley, and Stanford University's Luca Cavalli-Sforza. Wilson is most noted for his development of the "molecular clock," a foundational evolutionary dating tool that scientists use to establish divergence rates for human populations. Cavalli-Sforza was in the midst of collecting data for the most comprehensive books on cultural and molecular evolution, including the popular science book *Genes, Peoples, and Languages*. Though the two shared a similar evolutionary perspective, they had very different strategies for obtaining global representation.[31] Wilson was in favor of grid-sampling the globe—eschewing previous notions of relatedness and using geographical distance to test biodiversity. Cavalli-Sforza argued that grid-sampling was too expensive and risky, and that researchers would have to pay mind to predefined populations, marked by cultural or linguistic ties, in order to have a rough guide to biodiversity. In the end, planners agreed to collect immortalized, or reproducible, cell lines from four hundred predefined, relatively isolated populations while loosely distance-sampling nonimmortalized DNA.[32] Planners maintained that it was necessary to leave ethnic labels on the samples for future variation analyses. Therefore, unlike the Human Genome Project, the Diversity Project incorporated samples by ethnicity, individually labeled the cell lines with ethnolinguistic tags, and made DNA available on that basis. Neither of these protocols satisfied the new federal policies on race-based inclusion; however, they were in line with society's growing interest in diversity-conscious research.[33]

Cavalli-Sforza described to me the project's main intent: to gather a sample of the biodiversity that would have existed in the "population defined before 1492, before the great expansions." He insisted that the project was not so much looking for funding as for support from policymakers, the general public, and the U.S. government. From 1991 to 1994, volunteers from all over the world, mostly former students of Cavalli-Sforza, donated cell lines. As Cavalli-Sforza told me:

> I made two hundred. The Chinese ones were made by a friend of mine who was working with me. Pakistan donated another two hundred, and [that donor]

also was my student. Israel donated several. I didn't want any Jews, because I didn't want any anti-Semitic work going on, but they would have given them to me. But I took strict representatives of the Middle East . . . and again from France . . . and then all the others were from Ken Kidd from Yale.

Here diversity was defined as a collection that went beyond European descent, but not in the terms used in U.S. racial dialogues. Continuing to respond to outrage over geneticists' complicity in modern eugenics, these founders saw potential anti-Semitism as more of a concern than the racial implications the project would have for U.S. minority populations. Again, safety and practicality were envisioned in terms of avoiding certain groups—and their attendant ethical dilemmas—wholesale.

Despite these preoccupations, Diversity Project planners banked their project on its ability to overturn the colorblind, one-size-fits-all paradigm. They openly took issue with the no-labels approach of the Genome Project and criticized the exclusionary effects of colorblind sampling. Wilson and Cavalli-Sforza warned that the Genome Project's reliance on CEPH cell lines would result in a "Caucasian" or "Caucasoid" sequence.[34] The Diversity Project marketed itself as by contrast offering a globally representative approach based on self-identified ethnicity. By acknowledging the self-determined identity of tribes, the genomic venture claimed to be fighting Eurocentrism.[35] However, far from the federal government's vision, Diversity Project planners viewed inclusion on a global scale. In fact, project planners initially avoided contact with groups defined by "New World" histories of racial discrimination, in favor of groups thought to be "Old World" genetic populations. They reasoned that groups like African Americans were genetically uncertain, marked by recent admixture and continual gene flow, and not isolated enough to mark distant, consistent ancestry.[36]

In 1993, as the NIH completed the *Revitalization Act*, decisively tipping the balance in favor of a race-positive genomic science, Health and Human Services began inspections of genomic protocols.[37] The agency ordered the National Research Council to form an ethical commission to investigate the Diversity Project's inclusionary method. Stanford bioethicist Henry Greely, a leader in biotechnology law and ethics, was assigned to the project to formulate an ethical protocol that would address the recruitment of vulnerable populations.

Though the Diversity Project had paved the way beyond the colorblind paradigm, in this new climate its *race-free* sampling protocol had to be rethought.[38] As the project met with increasing suspicion from the science community, the general public, and indigenous groups around the world, the project intensified its claim to alleviate racism. Around the time when the infamous tract on IQ heritability, *The Bell Curve*,[39] was released to a wave of criticism, which included Diversity Project supporters, Cavalli-Sforza devoted one fourth of a UNESCO speech to discussing "how the Project will help combat the scourge of racism."[40]

Faced with pressure from social movements and the federal government, the Diversity Project eventually reassessed its ethnic, isolation-based sampling protocol and opened the project to U.S. racial minorities. However, as Reardon has argued, the initial stance of the planners was built into the sampling protocols of the project. The Model Ethical Protocol inclusionary policy that they later devised did not generate the fundamental shift to bring the project up to date with an increasingly race-positive emphasis in medicine and public health. Further, the project's focus on "minor ethnicity" foreclosed investigation into racial minority biostatistics.[41] Without mechanisms for reporting inclusion in U.S. Census terms, the project had little appeal to the federal government.

The Diversity Project was also criticized for not cultivating direct relationships with the groups they intended to sample. Thus, for many indigenous groups DNA sampling was seen as an imposition with little promise of return. Planners' refusal to face the history of race relations between Anglo-American academics and indigenous subjects prevented them from seeing their project as possibly implicated in that troubled history.[42] Indigenous groups around the world signed petitions and declared their unwillingness to participate in what became known as another instance of "biopiracy" and "modern-day colonialism."[43] Representatives of African American communities also protested the Diversity Project.[44]

After three years of minority lobbying against the project, despite the production of a revised sampling protocol based on obtaining community-based consent, the Diversity Project's major funding sources withdrew support. The National Research Council recommended that the Diversity Project restrict its scope to the United States so that the government could better monitor the project's community intervention strategies. The council's fundamental critique of the project was directed at its sampling protocol.[45] UNESCO's International Bioethics Committee warned that minorities and formerly colonized people were especially vulnerable to the potential for racist interpretations of project data.[46] At the close of the decade, the Diversity Project's race-free, ethnicity-based protocol had yet to find support from state and society; but a precedent for addressing racial matters with project goals was established. Genome projects became racial projects, projects dedicated to building racial equality with genomic tools. They also became conduits for the manifestation of interinstitutional civic platforms dedicated to making actionable meaning of human difference.

From Genome to Polymorphism, Diversity to Demography

Witnessing the Diversity Project's pending demise, planners at the Human Genome Project began work on an alternative, institutionally in-house human variation project that would directly incorporate race-based inclusion into its sampling protocol and bring genomics into line with the current racial paradigm. In 1996 the National Center for Human Genome Research and the Department of Energy released the *Guidance on Human Subjects Issues in Large-Scale DNA Sequencing*.[47] *Guidance* called for greater inclusion of women as subjects in genetic research. While it acknowledged that there were no male or female genomes, policymakers argued that "perceptions" of exclusion were significant enough to warrant a proactive inclusionary policy.[48] In other words, the social experience and political concerns about difference were to be a new part of the genomic research agenda whether they made sense or not at the level of the genome.

In 1996, ELSI also pursued a policy for intervening into minority communities by funding conferences on the significance of Genome Project knowledge for minority communities. A diverse array of scientists, social scientists, ethicists, officials, and community members was brought together to determine a model ethical protocol based on ongoing community partnerships and engagement. Genomics was moving toward investigation into the social and biological dimensions of race.

At a time when academia and government were beset by the racial politics of IQ studies, the commemoration of the Tuskegee syphilis experiments that had denied African American men life-saving treatment, and augmentation of Directive No. 15's political might, the National Human Genome Research Center was elevated to the status of a national institute.[49] The newly coined National Human Genome Research Institute was awarded the preeminent leadership role on matters of human genome variation and race. It moved immediately to create a genome project based on federal interests.

That project, the Polymorphism Discovery Resource, launched in 1997, was the first sequencing project dedicated to providing representation of the world's genomic diversity by Census race. Project planners saw the potential for researchers to make racist claims based on their racially sorted database, so instead of releasing the DNA in racial packets, they anonymized and pooled it into subsets of eight, twenty-two, forty-four, or ninety samples that reflected the proportional racial diversity of the total sample.[50] Discovery Resource planners explained to me the confusion that such a race-based project created for the field. Francis Collins, the present director of the NIH, was then the director of the Genome Research Institute and leader of the Genome Project and Discovery Resource. He remarked on the trouble the Discovery Resource had recruiting Native Americans in the wake of the Diversity Project and how community deliberations led to the project's disambiguated policy:

> At that point because we wished to include American Indians, and they felt very strongly that having samples identified as coming from Indian tribes was unacceptable, we made the decision that for that resource—which was really just there to discover variation at that point and not try to connect it to anything—that it was reasonable to have no labels.

Once again project planners calibrated their sampling protocols to the political landscape. Yet instead of outright avoiding vulnerable populations, scientists engaged leaders of minority communities in ethical dialogue from the outset. Collins and his colleagues said that through the process, scientists learned how to work with minority communities and better represent their needs. This included ceding scientific goals to community goals when contentions arose about group identification.[51]

Like the Human Genome Diversity Project, the Discovery Resource adopted a strong "redressing Eurocentrism" argument to popularize the project with the broader public. The project promised to at last "reflect the diversity of the human population."[52] In a report in *Science*, project representatives implied that its scientific rationale would generate social justice by bringing minorities into the fold of genomic research yet protecting their privacy:

> [P]ublicly available DNA collections contain little African material; Native American and Asian contributions are similarly scant. As a result, say NHGRI staffers, it will be essential to collect DNA from a racially structured set of donors. Once the DNA has been sampled, however, all personal and racial data

will have to be removed to protect privacy—diminishing the scientific value of the project, but bolstering its ethical foundation.[53]

In enlisting minority community leaders and joining the needs of minority access to genomics with those of genomics' access to minority samples, the Discovery Resource promoted a civic platform palatable to all. In its wake, the Genome Research Institute was viewed as an innovator of federal inclusionary efforts.

By the time Craig Venter and his company, Celera, entered the race to map the human genome, in 1998, questions of race had become paramount.[54] A former NIH biochemist and researcher, Venter is now known for his work in cataloguing the world's oceanic diversity and for being the first to create a synthetic living organism. Venter told me that representing the range of human diversity was critical for him from the very beginning. Although it was a struggle to get board approval to study all the groups he desired to, he felt that he was successful in covering the major axes of diversity, in terms of both gender and race:

> Many on the committee were very concerned that the data would be used to justify racism for the same example I gave you: take any two people; they are going to look different, and therefore the two races are different, right? . . .
>
> But, I felt it was important *symbolically* for the first "human genome" to [be representative]. . . . So, we made decisions on a pragmatic level of DNA to cover, but out of the pool of volunteers we tried to select people to have as much sexual and ethnogeographic diversity as we could in five people.

Venter's remarks show that by the end of the decade, race, diversity, population representation, and global comprehensiveness were considered *at the moment of project design*.[55] Project leaders had internalized the goal of creating diverse representation with their research, and many were approaching genomic sampling in terms of a political project worth fighting for. Venter negotiated a protocol that squared with his own antiracist values and the dominant framework of inclusion. This protocol of picking subjects of different morphologies depended on a strong consciousness of racial difference.

With his committee Venter also deliberated the merits of representation in light of potential misuse of genomic data on minorities. In fact, the team faced pragmatic issues of technical acuity by weighing the potential social outcomes of using different measures.[56] They made the decision to include diverse racial groups rather than shield minority groups from the potential for racist scientists to exploit their DNA. Though the project was restricted by feasibility and cost, thus limiting the final sample to five people, the team made sure to represent as many populations as possible; they were convinced that variation was the necessary focus of genomics and that race-conscious sampling was superior to colorblind sampling.

Policy Loops

From the end of the Genome Project to the present, we see feedback loops in policy and protocol that bear the marks of an emergent racial paradigm based on a combination of social and biological inquiry.[57] In the Genome Project's final Five Year Plan, "New Goals for the U.S. Human Genome Project: 1998–2003," project leaders called genomic variation "the fundamental raw material for evolution" and claimed that its analysis raised "additional ethical, legal, and social issues that need to be anticipated, considered, and resolved."[58] The plan's ELSI goals explicitly promised to "Explore how socioeconomic factors and concepts of race and ethnicity influence the use, understanding, and interpretation of genetic information, the utilization of genetic services, and the development of policy."[59] Following the constructionist turn in racial scholarship and politics, race was conceived of as a social fact that mediated one's access to genomic information and services.

ELSI institutionalized the field's new orientation to race by circulating a request for applications entitled "Concepts of Race, Ethnicity, and Culture: Examination of the Ways in Which the Discovery of DNA Polymorphisms May Interact with Current Concepts of Race, Ethnicity and Culture." Questions from the 1999 call included:

> How will individuals and groups respond to potential challenges to or validations of their racial, ethnic or cultural self-identification, based on new genetic information?

> What impact will the discovery of information on genetic variation have on the current categories (as defined by the government) of race and ethnicity?

> Will the discovery of patterns of DNA polymorphisms affect the concepts of race and ethnicity that are used by genetic, anthropological and other health and social sciences researchers?[60]

For the first time, scientists were directed to question how their research was going to affect racial *beliefs* and *identities*. Behind closed doors, the institute's National Advisory Council also discussed the conceptual structure of race and ethnicity. At this time, institute scientists debated the creation of a permanent in-house interdisciplinary Race and Genetics Working Group composed of natural and social science experts.

President Bill Clinton famously remarked in June 2000 that the completion of a draft sequence of the human genome proved "in genetic terms, all human beings, regardless of race, are more than 99.9 percent the same."[61] His assumption that scientific practice had something powerful to contribute to solving the sociopolitical conundrum of race showed just how much genomics and the social reality of race were being coenvisioned in the public sphere. Academic affirmative action was again coming under attack, and politicians of all persuasions sought a permanent, irrefutable solution to identity politics in the biology of race. At the same time, genomicists wanted to send the message that genomics was serious about inclusion *and* antiracism through a leveling of the social and biological playing fields.

Clinton's remark hailed genomics as a political and scientific arbiter of the meaning of race.[62] At a critical juncture when genomics finally came into its own institutions, journals, professional associations, and funding streams, it was given one of the highest displays of support for its interest in race. The public face of the genomic field would now be shaped by its ability to approve or deny particular notions of race. Genomicists initially did not want to deal with race; yet, set as they were between a genetics with a harried past on race and a general public interested in new knowledge about race, scientists were forced to articulate their own beliefs.

With Clinton at his side, Venter rejoiced: "I am happy today that the only race we are talking about is the human race." Venter claimed that it was impossible to disambiguate the five major groups present in the Genome Project's sample; once translated into genomic data, race disappeared.[63] In the popular media, a *New York Times* article quoted directors Eric Lander and Aravinda Chakravarti as calling race "a bogus idea."[64] In the *Boston Globe*, Lander echoed: "The whole notion of differences between us has been about ranking people.... We need to get that out of our heads." Venter's assertion that "No serious scholar in this field now considers race to be a scientific concept" indicated—ironically, through refutation—that race had become a salient talking point for the field's representatives and a prism for the interpretation of genomic advances.[65]

Meanwhile, pharmacologists were developing methods for understanding the human variation of drug response. The joining of market-based and academic knowledge spawned a pharmacogenomics of race.[66] After the FDA issued the Demographic Rule, it was common to see clinical trials refer to the variable drug reception of "race-sex" or "race-gender" groups, or for reports to begin by articulating the influence of "race and ethnicity" in drug metabolism. Within genomics, too, a growing body of research claimed that pharmacogenomic racial disparities had to be further explored. Variants in certain gene families had been shown to differentially affect reactions to a range of lifesaving therapies used in everything from cancer to postoperative care.[67] Findings like these suggested a sea change in medical treatment and drug development, wherein a new system of personalized medicine based on pharmacogenomic diagnostics would take hold. Genaissance Pharmaceuticals CEO Gualberto Ruaño contended that drugmakers could target the racial groups known to benefit from a drug: "Efficacy could be proven in small cohorts of a few hundred patients compared to 3,000 or 5,000 as we do now."[68]

In 2001 the *New England Journal of Medicine* published the two main articles that set Jay Cohn and researchers of the African American Heart Failure Trial on course to produce BiDil. One of the studies, on a common antihypertensive angiotensin converting enzyme (ACE) inhibitor, made waves throughout the research community when it reported a lesser response in "black as compared with white patients" with a specific kind of heart failure.[69] The other argued that an alternate beta-blocker formula provided benefits "of a similar magnitude in both black and non-black patients with heart failure."[70] Though these studies did not reference genomic data—they merely conventionally reported the differential effects of two pharmaceutical therapies on people who self-identified as "black" or "white"—they were nevertheless taken up by the science and news media as proving a genetic basis for racial difference that was manifest in health outcome.

The articles were accompanied by two editorials that called on genomics to mediate the study's results.[71] One, written by deputy editor Robert Schwartz, called the researchers' use of race outmoded: "The Human Genome Project now gives us the power to uncover the true origins of genetic variations; linking them to race has become passé."[72] Schwartz encouraged biomedical journals to create policies that would eliminate the use of race outright, and to provide a richer definition of other classificatory terms. The other editorial, written by drug therapy editor Alastair Wood, supported a genetic basis for differential drug response and race-based medicine, yet expressed the need for scientists to locate the gene target or "genetic determinants of the racial differences, rather than [put] attention to the external phenotypic manifestations of race."[73] Wood had himself recently conducted a series of studies into the racial and ethnic pharmacogenomics of beta-blockers and was concerned that the present research had not identified the genetic causal factors. Though these editors disagreed as to whether there was enough of a genetic basis for race as to warrant its use in genomic research, both credited the human genome as providing the ultimate solution to society's racial quandary. With Cohn and his team of BiDil researchers, they called for more research on minorities and a more assertive inclusionary policy in biomedicine. Yet they also insisted on more genomic research into the "real reason"[74] for disparities.[75]

This journal issue set off a frenzy in the major news media, which opened the floor to more elite stakeholders. When the *New York Times* asked, "Shouldn't a Pill Be Colorblind?", the National Human Genome Center's Georgia Dunston told *CNN Live at Daybreak*: "When we look at DNA, there is no way that you can group, organize, sort genes in such a way that you've got some collection of genes that would correspond to the racial groups that we use."[76] Evolutionary geneticist Joseph Graves told *NPR* that races were too genetically mixed to be typed for drugs.[77] Venter replied: "It is disturbing to see reputable scientists and physicians even categorizing things in terms of race. . . . There is no basis in the genetic code for race." Yet Cohn countered: "Here we have the black community accepting the concept that African Americans need to be studied as a group, and then we have the science community claiming that race is dead. . . . It seems to me absolutely ludicrous to suggest that this prominent characteristic that we all recognize when we look at people should not be looked at."[78] Cohn appealed to "the man on the street" to support the idea that race was a reality worth exploring.[79]

Smaller scandals, such as James Watson's claim that there was a genetic basis for racial differences in libido and in body weight ("That's why you have Latin lovers. . . . You've never heard of an English lover—only an English patient"[80]), also threw the field into contention. But as scientists defended their "antirace" stance against public health officials, physicians, and policymakers, their unified message began to fracture. Intramural debates of this nature would eventually lead scientists to flesh out their understandings of the social implications of race and take a greater public responsibility for the meaning of race.

In July 2001 the National Human Genome Research Institute held a two-day conference on global population sampling and its scientific and social implications to hammer out the field's HapMap project. At this meeting, planners tackled the merits of continent-based sampling versus minority inclusion by

Census race, and the potential for genomic research to cause racial stigmatization.[81] While these scientists worried that a continentally structured genome project would send a message that distinct racial boundaries existed, they agreed that understanding global variation required a proactive inclusionary protocol. As Lander argued: "We must make sure the information is not used to stigmatize populations. But we have an affirmative responsibility to ensure that what is learned will be useful for all populations. If we shy away and don't record the data for certain populations, we can't be sure to serve those populations medically."[82]

Across the field, the assumption that race had no place in biomedicine was rapidly giving way to a keen interest in a genomic investigation of racial health disparities. The Census policy was beginning to take hold across genomics' central funding agencies, and policy analysts were calling for greater attention to genetic causes for health disparities.[83] The NIH set to launching a $33 million study of heart disease and a $22 million study of cancer in African Americans.[84] Both of these cohort studies set racial variation as a primary interest.[85]

The field's notoriety would thereafter hinge on its management of the race controversy. Whether delivering the message that genomics had proved race to be biologically unsound or finding new ways to incorporate racial groups into research, scientists would speak as the authority on race. Offering new knowledge about race gained the field some autonomy from the political field of Congressional deliberations, and from other scientific fields like population genetics and epidemiology that had previously cast race in biological terms. Genomic arguments about race, in fact, placed the field in a superior position to these other fields, as its socially antiracist yet biologically anchored findings have tackled head-on the biologically bereft federal taxonomy as well as the racist biology of the past.

Following the White House convocation, the NIH released a new genetic sampling policy modeled after ELSI recommendations to require minority community consultation on all studies;[86] in accordance with Health and Human Services' newly formulated Initiative to Eliminate Racial and Ethnic Disparities and *Healthy People 2010*,[87] the agency drafted a Strategic Research Plan to Reduce and Ultimately Eliminate Health Disparities, committing $1.3 billion in funding.[88] In the transinstitute strategic plan as well as the mission-specific strategic plans of individual institutes, the integration of genomic research figured as a foremost tool in the fight against racial health disparities.[89] Plans began with lists of diseases known to disproportionately affect minorities, and were followed by notes on the genomic studies that would be supported to address those diseases. They concertedly argued that racial inclusion needed to be built into the earliest stages of translational medicine, thus stressing the importance of basic biological research to their goals. The transinstitute plan also made provisions for minority research awards and training programs—including an ELSI career award and a Genomics Short Course for Faculty at Minority Institutions—and racially specific public input sessions on genomic research policy. The Genome Research Institute was exemplary in this regard. It devoted nearly its entire strategic plan to preparing for a minority-led disparities revolution within the field. Of the plan's fourteen initiatives, twelve focused on recruiting minorities into scientific roles and decision-making processes. Again the institute sought to bring broader issues of educational and socioeconomic disparities, access to genomic knowledge, and racial participation to the center of genomics. With a race-positive policy for research inclusion as well as *researcher* inclusion—a policy akin to affirmative action in the broader social sphere—the paradigm shift was now complete.

Converging Interests Within the Triple Helix

While the Human Genome Project continued to complete its sequence, through 2003, the Genome Research Institute was already planning its next global project. The Discovery Resource, with its inclusive yet disambiguated database, was soon eclipsed by The SNP Consortium, a race-positive sequencing project engineered by thirteen major pharmaceutical, genomics, and informatics companies and the Genome Project's British sponsor, the Wellcome Trust.[90] The SNP Consortium, founded in April 1999, aimed to generate a publicly available map of three hundred thousand evenly spaced SNPs (0.001 percent of humanity's variation). In July 2000 the Genome Research Institute donated twenty-four Discovery Resource samples "representing several racial groups"[91] and provided database and storage infrastructure to promote the generation of 250,000 additional SNPs.[92] Within months, the consortium had produced a SNP map detailing 1.42 million common SNPs.[93] The consortium also confirmed that in accord with principles of genetic recombination, SNPs were inherited in haplotypic segments; thus, knowing one SNP meant that a succession of flanking SNPs was also known. Haplotypes could serve as a shortcut to determining the total SNP variation in the human genome. Scientists at Whitehead Institute/ MIT Center for Genome Research—the academic precursor to the Broad Institute—had just developed a new haplotype mapping technology and applied it in two global survey studies. In August 2001 the Genome Research Institute announced its plan to expand the Genome Project and The SNP Consortium into a global and globally representative haplotype sequencing project.[94] Academic, governmental, and market treatment of race converged triple-helically towards another Census-based model of human diversity.

Unlike the Genome Project or the Diversity Project, from the outset explicit racial interpretations framed the project. In a preproject analysis of the haplotypes on chromosome 21, Perlegen found that comparing consortium samples to twenty-one privately owned samples elicited even larger segments of inherited variation. Perlegen spoke of their own samples as "representing African, Asian, and Caucasian chromosomes" and claimed "more than eighty percent of a global human sample can typically be characterized by only three common haplotypes."[95] Sample data were submitted online by Census race.[96] This conflation of continental variation, global population representation, and federal standards was in line with the Discovery Resource framework and the field's move to deliver basic research messages in a racialized form.[97] Genaissance also reported a global survey of haplotype variation in "twenty-one whites, twenty blacks, twenty people of Asian descent, eighteen Latinos, and three American Indians," finding important gene frequency variations between the groups.[98] The SNP Consortium continued to use the U.S. federal government's Census taxonomy to create a map of the most prevalent molecular variants in the human genome. A study of variation between the consortium's Northern European and Nigerian sample populations proved that racial comparison could elicit key human migration patterns, such as population bottlenecks responsible for entire continents' genetic variation. Together these studies buttressed the notion that shuttling between Census categories, widely accepted lay race categories, and continental variation was a viable practical approach for the field. Moreover, they supported the stratification of biological populations by commonsense lay race. Subsequent to these publications, Perlegen was welcomed by the Genome Research Institute to be a primary sequencing center for the International HapMap Project.[99]

Lynn Jorde is the populations and ELSI cochair of HapMap and president of the American Society of Human Genetics. His research on topics like mobile elements, SNP tagging, and whole genome sequencing in families has kept him at the forefront of genomic methodology. Jorde explained the intensity of dialogues about race during the early planning stages of the HapMap Project.[100]

Project planners were especially fearful that the project would fumble its use of race, as the Diversity Project had done:

> There was a real fear though that the whole thing would get hijacked the way HGDP was; and get turned into a political controversy, and for that reason, all it takes is a few people in Congress to say, "Wait, why are they doing this in the first place?" And then the whole thing just falls apart. . . .
>
> Do we need a complete sampling of human diversity? No, we don't need to look for that. So, let's get what we need to better understand the haplotype structure of the human genome as expediently and as politically safely as possible.

Jorde said the project planners were concerned that HapMap would be seen as "HGDP II," and to avoid that outcome they constructed the project "in such a way that it [didn't] get a bunch of groups upset." Again political safety was a concern for scientists; however, this time they saw safety as a matter of careful racial inclusion and representation. Groups that had been considered off-limits in the Diversity Project, such as Jews, were here considered for inclusion. HapMap planners took pains from the outset of the project to anticipate how these goals would be achieved. Planners were acutely aware that government standards and public values needed to be weighed at the same time as views of minority communities and the broader public, who would have to be enrolled on their terms.

Eric Lander—cochair of President Barack Obama's Council of Advisors on Science and Technology; a founding director of the Broad Institute, Millennium Pharmaceuticals, and Infinity Pharmaceuticals; professor at MIT and Harvard; and a leader of the Human Genome Project—reflected on these matters in the context of the project's decision to sample the DNA of one population from each of the main continents. Speaking as the lead investigator and chair of the HapMap Methods Group, he explained:

> We know that there are regional similarities. We know that in Africa there is somewhat more variation, because all mankind comes from Africa. And so in order to get some idea whether regional variation mattered, it was useful to pick a site in Africa, a site in Europe, and a site of Asia. If one simply picked with a worldwide grid sample, you wouldn't find out.

To ensure representation of global diversity, planners used ethnic communities as proxies for shared continental ancestry. Referring to the rejection of grid sampling, many like Lander further distanced the HapMap from the Diversity Project. Planners emphasized using a sampling method that would allow scientists to form lasting *political* relationships with the people they sampled. Participants were enrolled through a lengthy community consultation process that was marked by certificates, ceremonies, and cultural celebration.[101]

The original proposition to sample from the main continents was altered when representatives from China and Japan expressed their intent to participate. Both China and Japan were included, yet their sample contributions were each halved. Pui Yan Kwok, head of sequencing at the University of California, San Francisco, branch, said that groups who supported the project both financially and ethically were enrolled. As a steering committee member of the International HapMap Consortium and a longtime developer of DNA sequencing technology, Kwok is one of the project's foremost U.S.-based Chinese researchers. He is also a founder of the Human Variome Project—a global initiative to produce a centralized variation-genomics database sponsored by organizations like UNESCO, the European Commission, the World Health Organization, and the CDC. He described the enrollment process as one where project representatives "did not ask everybody in China" but rather targeted local elites in different "centers" of Beijing: "There were kind of discussions in the public—not really marketing, but there was public discussion about it. The Chinese were very, very enthusiastic about it." For town meetings, U.S.-based recruiters whose ethnicity matched the sample population's were chosen to represent the project, while esteemed community leaders were chosen to represent the desired research population. Individual researchers were thus called upon for their cultural and political expertise to create a synergy of proxies between project, community, and continent. This strategic essentialism, a tactic imported from the mainstream arena of racial activism, would become a lynchpin of genomics' way of managing knowledge about race, throughout the HapMap Project and into the present day.

The original race-based protocol was altered a second time when indigenous groups voiced their opposition to participating in the project. These groups requested that HapMap not include indigenous or aboriginal samples. When I asked Lander why Native Americans were not included, he replied:

> This portion was political and appropriately so. That there are significant Native American groups, RAFI, and others who felt that the potential for exploitation was great—and historically it *has* been significant. And getting the community in a somewhat broader group of consent was going to be a lot harder.
>
> And so I think it was important, it was a pragmatic and appropriate measure,

to choose regional groups where there was also some reasonable prospect of getting consent. Choosing groups that have been subject to discrimination in the past introduces a whole other level of complication.

Kwok, in reply, was more frank:

> KWOK: That's actually political. At the beginning, when we were talking about the project, the Native Americans were invited to these kinds of workshops, and there they basically told us nobody was—I think what they said was they just did not think that they [the project findings] would benefit them necessarily. Number one, they are giving us something, but we are not giving anything back. And the other is that they don't like the conclusions that we are going to draw from the data. I think they were mostly—
>
> BLISS: About evolution or something?
>
> KWOK: Yes, they were also unhappy with our telling them that they came from Asia, because they have their own traditions about how they came about. For those reasons, they do not want to participate.

Lander's remark about RAFI (Rural Advancement Foundation International), an NGO that led the bulk of the anti–Diversity Project campaign, again shows how the Diversity Project's failure galvanized the field and brought its elites into further consensus on the question of race. Scientists had become wary of the political implications of colorblind research. Though Lander implied that the project sought partnership with groups that could easily be enrolled, Kwok's remarks show that project scientists actually conducted town meetings with Native American groups. They were not able to convince members of those groups that the project would benefit them, or that knowledge from the project would not counter their own cosmologies. However, an elaborate recruitment process was made nonetheless, and scientists conceded to these representatives' demands.

Some scientists were critical of the resulting sample base. One scientist, who asked to remain nameless, said:

> So to me, it is quite a shame that Native Americans were not included in the HapMap from the beginning. . . . I am not going to be naïve here, but I think, given the Native American contribution to the U.S. gene pool, the U.S. resident gene pool, . . . [the] West African could be the second largest contribution in terms of genetics to the gene pool, but Native American would be right behind that.

Thinking in both genetic and demographic terms, this researcher fused minority inclusion with genetic inclusion. He also argued that downstream benefits, such as pharmacogenomics and diagnostics, were a chief concern. If Native Americans weren't included in the study, how would scientists later be able to use HapMap data to create personalized medicine for them?

From this perspective, planners trained a critical eye on the recalcitrance of community representatives. One said:

> Yes, it's a shame. I have different views about some of these things, but I think that sometimes the protectors of communities, or sometimes they are called gatekeepers, may not adequately represent those communities very well. I think Native Americans have all the right in the world to be very suspicious of certain tendencies and behavior. But I also think that it is equally important to make sure that you are not excluded in very important international endeavors.
>
> Again, we have to be careful because we may see that it is important, but other groups, they do not see that as important. But given the way that things like HapMap are driving our medical research right now, and may continue to do so in the future, I really think that we need to make sure that we have enough information for all human populations.

This scientist views access to cutting-edge medicine as a human right, just as clean water and sufficient shelter are. He worries that his colleagues did not select the appropriate community representatives—people who could appreciate a growing "genomic divide." This bioethical twist to the equation of inclusion and equality shows that scientists have forged evermore complex ideas about the political ramifications of genomic research.

When it came to collecting African samples, HapMap planners went with investigator Charles Rotimi's ties in his native Nigeria. Rotimi was widely respected for his leadership of a genetic epidemiology unit focused on diabetes in people of African descent. Today he is the director of the NIH National Center for Research on Genomics and Global Health; president of the African Society on Human Genetics; and leader of the first international African genome project, the Human Heredity and Health in Africa Project. Rotimi was strategically selected for his firsthand knowledge of the social and political structures with which HapMap would be dealing. HapMap scientists felt that they would have a better time of enrolling people in Africa, and of maintaining positive, ongoing ties with their sample base, with Rotimi at the helm.

David Altshuler, a founding member of the Broad Institute who worked closely with Rotimi, was lead investigator of The SNP Consortium before heading up HapMap's sequencing and analysis at the Whitehead Institute/MIT Center for Genome Research. He explained:

[The sample] wasn't representative. It was sampled from—*arbitrarily* sampled from. It gets to a very practical set of questions, which is like, *where*? Why Ibadan, Nigeria, instead of somewhere else in Africa? Purely practical. Because there were people who were connected to the research field—like Charles Rotimi—who were connected to the research field, who had been working there.

There was a discussion about [whom] should we sample—we could've done it all in America, and that was felt to not be a good idea. That for political, more than scientific reasons, that if you were going to have a sample from Japan, it would be good if it was collected in Japan as opposed to first-generation Japanese Americans. It's sort of a self-determination thing, you know, that Charles—somebody who is African and worked in Africa for many years—do it, as opposed to someone else.

A decade into the minority inclusion brought about by the *Revitalization Act*, Altshuler's comments show that genomicists were thinking race-positively and were well versed in notions of self-determination and strategic essentialism. They were able to justify the HapMap's equal parts European, African, and Asian DNA pool with the language of minority justice and racial equality.[102]

HapMap's official introductory article in *Nature* provided a now common inclusionary rationale for its protocol. It explained its sampling in terms of the project's ability to attain both biological, global representation and social inclusivity—that is, "inclusion-and-difference," and social and biological concerns for health. It called the Nigerian, Japanese, Chinese, European, and American participants representative of "four large populations [that] will include a substantial amount of the genetic variation found in all populations throughout the world."[103] The project's continental overtones moved the field a step away from the race-based medicine debates that were raging in the media, while allowing the project to continue to promise comprehensive benefits for minority groups.

Finally, anticipating the racial comparisons that have indeed flowed from publication of the HapMap, planners generated a set of *Guidelines for Referring to the HapMap Populations in Publications and Presentations* that instructs all users of HapMap data to adhere to location-specific descriptors: Yoruba in Ibadan, Nigeria; Japanese in Tokyo, Japan; Han Chinese in Beijing, China; and

CEPH.[104] HapMap no longer officially claims it has provided comprehensive global representation. *Guidelines* reasons that location-specific labels are more scientifically correct, but also politically correct. HapMap appeals to its users, when constructing their protocols and arguments, to show respect for the communities sampled.[105]

Engendering Self-Determination

The path from Discovery Resource to HapMap shows that political representation became central to constructing new genomic protocols. Whom to send into the field, to represent both global genomics and local participants, preoccupied project scientists and the many institutions within which they worked. Engaging the public was especially challenging outside the United States as scientists attempted to read local concerns through American optics.[106] In one instance of HapMap's first phase, Japanese representatives requested their samples to be referred to as "Asian," even though Chinese participants were to be labeled "Chinese." Project scientists found themselves in the position of denying participants their self-determined representation.

As HapMap got under way, a group of African American scientists attempted to correct this problem by conducting research into African American health in a "by the people, for the people" manner. Fashioning themselves as representatives of African American interests at the community level, genomicists at Howard University's National Human Genome Center ("a comprehensive resource for genomic research on African Americans and other African Diaspora populations, distinguished by a diverse social context for framing biology as well as the ethical, legal, and social implications of knowledge gained from the human genome project and research on genome variation"[107]) drew on their social positions and social capital to bring African Americans into the research fold and to prioritize the problems that African Americans face. Attempting to put racial health disparities on every lab's radar, these scientists shuttled the race-positive stance across private and public barriers.

In 2003 the National Human Genome Center held an interdisciplinary, transinstitutional conference entitled "Human Genome Variation and 'Race.'" With funds from the Genome Programs of the U.S. Department of Energy, the National Human Genome Research Institute, the Irving Harris Foundation, and Howard University, the center convened the field's leading experts to discuss what genomics could say about race, what was still unknown, what research was needed, how data could be applied to benefit people, how it could hurt people, and finally what genomicists needed to do to promote "beneficial social outcomes."[108] The conference proved to be a turning point, as elites with contrary views met in order to find appropriate uses of genomic technologies and concepts in the search for understanding racial difference. These scientists agreed to disagree on the utility of race for structuring genomic studies, but they concurred that genomics would have vital answers about race for the general public.

Large-scale federal investment soon followed. In the case of the National Human Genome Center, the Genome Research Institute was on the verge of issuing a new funding mechanism for ELSI research, and Howard, with its role in the black community, would be a part of the first cohort.[109] The Centers of Excellence in ELSI Research program was a $3 million commitment to fund leaders in genomic bioethics. Policymakers wrote technology development and partnership with minority institutions into the research funding announcement. Sickle cell disease research was to be a first focus of the new program. They also provided funding for a pool of minority graduate students. In sum, the Genome Research Institute's three-pronged initiative combined: support for minority institutions and investigators focused on health disparities research, recruitment of minorities to train in genomics, and technology development to be able to study minority health with genomics.

With this concerted support for race-focused research, the National Human Genome Center announced plans to form an African American Diversity Project—the first race-based genome project. The plan was to sample twenty-five thousand Howard Hospital visitors over five years and create a DNA database, or biobank, that would generate more research on African Americans. Spokesperson Dunston ensured the public that the bank would be filling an important gap in knowledge about African American health. From her position as a pioneer of African American population genetics and founding director of the center, she argued: "This knowledge is critical, and there is no substitute for participation."[110] America's leading African American medical associations unanimously supported the project. They saw genomics as a social good to which African Americans had had little access. Rotimi, then a director at the center, framed the biobank's main purpose as extending personalized medicine to African Americans: "If you want your clothes to fit, you'd better go to the tailor to be measured."[111] With the Icelandic biobank, housed by DeCODE genetics, this would be the second large-scale population-based biobank in existence and the only biobank attuned to the health effects of racism.

Although the project did not raise the funds it needed in order to come to full fruition, the center's scientists had decidedly shifted the terms of genomic research toward race. These scientists were able to make the case for an explicit racial project because they were known to be affiliated with the community of beneficiaries. Furthermore, they possessed the personal identity capital that other genomicists lacked. Amid the broader social shift toward race-positivity and the shift within genomics from experimental research to translational medicine, it was thought too that minorities' wariness about being vulnerable to exploitation as experimental subjects would change into concern about being excluded from a potentially enormous technoscientific advance.

On November 11, 2004, in the *New England Journal of Medicine*, Cohn, Anne Taylor, and the African-American Heart Failure Trial Investigators reported the successful results of BiDil's final clinical trial. Months before, the researchers had abruptly ended the trial when they saw a 43 percent relative one-year mortality decrease in African American subjects taking the drug. The researchers admitted that they had no idea what caused the drug to work, but they linked its success to the unique pathophysiology of African Americans. While the researchers supposed that a future study of "genotypic and phenotypic characteristics that would transcend racial or ethnic categories" would be fruitful, they strongly encouraged the pharmaceutical industry to use their nongenomic, race-based study as a model for rationalized medicine.[112] They warned, "the average effects in heterogeneous populations may obscure therapeutic efficiency in some subgroups and the lack of such efficacy in others."[113] The researchers maintained that targeted medicine should be the new paradigm in medicine.[114] Backers such as the Association of Black Cardiologists and the National Medical Association called on genomics to support race-based medicine and incorporate race-specific efforts into their pharmacogenomics research.[115]

The following February, Perlegen released a cluster analysis of haplotypic variation, and in October the HapMap celebrated Phase I completion.[116] Immediately published online in a public database and made available through a national biobank, its reference samples became the new standard for genomic research. Genomicists revisited earlier methods with a new haplotypic lens[117] and began searching for disease risks associated with the possession of certain haplotypes.[118] The hunt for signs of natural selection and other evolutionary reasons for the variation that now exists not only became a possibility but a focus of the field.[119] With their continental infrastructure, these databases also became the newest source of racial claims.

Of late there has been widespread support for racial technologies and therapeutics coming from funding and regulatory agencies, health justice organizations, and representatives of minority communities. Though BiDil has been unsuccessful on the open market, institutions like the American College of Medical Genetics and the FDA have recently called for race-based testing of drug response blockbusters. These organizations do not recommend the diagnostic for scientific efficiency but rather, as Jonathan Kahn puts it, "to perform a sort of economic triage to focus on those for whom the test is most likely to produce a useful result."[120] Institutions and advocacy groups see these shortcut diagnostics as a way to incentivize the study of minority health.[121]

The Association of Black Cardiologists has backed too the production of Bystolic, the beta-blocker designed for hypertension therapy in African Americans. Ironically, Bystolic is marketed as a corrective to the *underuse* of hypertension therapy in African Americans resulting from the success of the BiDil campaigns in convincing African Americans that beta-blockers and ACE inhibitors work less well for them.[122] The president of the Association of Black Cardiologists, Paul Underwood, has shown keen acceptance of the race-based premise, calling Bystolic "another therapeutic tool to [add to] the armamentarium in the treatment of high blood pressure in African-Americans." African American advocacy groups like the Association of Black Cardiologists and the National Medical Association have petitioned insurance companies to add these drugs to their formularies and to encourage doctors to support them. Thus, even though these clinical research studies showing "race-specific efficacy" are carried out by scientists outside of genomics and without the search for genetic variants, their political and financial support is a mainstay of the pharmacogenomics of race.

Diversity's Future

In only twenty-five years the institutions that have defined genomics have moved from practicing colorblind science to race-positive inquiry. Not only have the field's elite ramped up their involvement in racial debates; they have brought the field in line with a new civic platform that values the consideration, if not outright use, of racial categories derived from the political field. Genomicists have attempted to reformulate the content of policies and the makeup of the genomics workforce in light of minority inclusion goals. In doing so, they have appropriated the logic of racial activism from the broader public sphere.

Yet inclusion isn't the only basis of the new paradigm for understanding race: a dual interest in social and biological factors is of chief concern. The new language of race is spoken in terms of genetic and environmental factors, while concepts of equality hinge on promises of scientific and social-structural change.

Whether immersed in a world dominated by colorblindness or race-consciousness, these events show that normative synergy happens when institutions collaborate across the social domains of academia, government, industry, and the nonprofit sector. Struggles for scientific autonomy and authority are also marked by the normative effervescence between institutions and the public sphere. Genomics has reached its place at the apex of the biosciences in part because it has consistently shifted to meet the dominant American racial paradigm. In fact, racial projects that were unthinkable in earlier years are now some of the most high-profile projects around, because they are consistent with the new dominant paradigm.

Two success stories from the end of the decade of the genome are of note. First, the National Genographic Project, a five-year study that collected the mitochondrial and Y-chromosomal DNA of hundreds of thousands of people, completed its first phase in 2010. This study of "indigenous and traditional populations and the general public . . . to reveal man's migratory history and to better understand the connections and differences that make up humankind"[123] differs from the earlier Diversity Project only in its use of nonimmortalized DNA. Though RAFI and a number of indigenous-peoples councils protested the project at its launch, by 2006 the criticism had all but ceased.[124] The Genographic Project donates the proceeds of its test kit sales to a legacy fund for indigenous cultural revitalization. The project's founder, National Geographic's explorer-in-residence Spencer Wells, has authored a number of popular books on race and human variation based on this research.

Second, the NIH and the Wellcome Trust have launched the Human Heredity and Health in Africa Project, a $25 million, five-year gene-environment analysis of common diseases affecting sub-Saharan populations. This is the first project dedicated to a single continental group. Though it has been rationalized in the same way as the African American Diversity Project, it has found full financial support in the global market. As its director, Rotimi, stated at the project's first press conference, "we would like to make sure that whatever benefit that we're going to accrue from using genomics to understand health and human history that [it] doesn't go past Africa, as other revolutions have done in the past."[125] The project banks on its ability to include Africans while recruiting local scientists and delivering targeted research. Diversity, equality, and intellectual development through genomics is the field's governing norm.

Indeed the major genomic projects are now framed and evaluated in terms of their costs and benefits to minority groups. Still we may wonder, can a race-positive genomics bring about racial equality? Racial projects have typically been waged by collectives formed on the basis of shared life circumstances of racial privilege or inequality. Those circumstances have political genealogies linking back to colonization, enslavement, exclusion, and immigration. There is no doubt that genomic racial projects are political acts; but can biology-based projects explain these histories?

As policies change and public sentiment about what race should mean fluctuates, the field will continue to sharpen its inclusionary framework. Yet the issues of who can speak for whom, and what the right kind of support is for racial therapies and technologies, go unresolved. When the Association of Black Cardiologists contributes to a study for race-based medicine, should the field stand behind it? Which institutions are best qualified to represent local communities? These issues also open up questions about the *enactment* of policy and protocol. In their journey from color-blind to race-positive science, genomicists have become adept at thinking with racial implications in mind at the start of research. It is thus important to further explore how genomicists bring questions of race into sharper focus through the values, concepts, and practices that they hold in everyday research.

3   The Sociogenomic Paradigm

I N J U N E 2003 police officers tracking a Louisiana serial killer used a racial profile constructed from DNA left at the scene of the crime to catch the perpetrator. Louisiana law enforcement had been searching for a white male suspect after having been tipped off by witnesses to the crimes. However, the results of a DNA test provided by Florida-based DNAPrint Genomics, Inc., indicated that the suspect was a man with "eighty-five percent African ancestry and fifteen percent Native American ancestry."[1] The chief scientific officer of DNAPrint, biochemist Tony Frudakis, told the officers, "You're wasting your time dragneting Caucasians; your killer is African American." Mark Shriver, inventor of the test and other admixture mapping software based on ancestry informative markers, insisted that genomic admixture studies proved that lay stereotypes were incongruent with genetic ancestry. *Good Morning America* quoted him as saying that the test dispelled the notion that all serial killers are white.[2] When law enforcement officials apprehended Derrick Lee Todd, an African American man who turned out to be responsible for the killings, Frudakis said the technology was like a more reliable eyewitness account, one that could assess a person's racial proportions.[3] At the time, the company was busy developing tests based on markers associated with skin color and eye color.[4] Police departments began signing up to be the next to successfully apply the technology.[5]

DNAPrint's marketing of racial "PhotoFit" technology created a firestorm inside and outside the field of genomics. A number of ELSI bioethicists warned that these technologies would lead the public to believe that races were distinct genetic populations.[6] But genomicists had been developing the admixture technology at the base of these applications for as long as the field had existed,[7] and that technology was already being used in disease studies ranging from asthma and cancer to the pharmacogenomics of alcohol and drugs.[8] Admixture mapping was the tool of choice for scientists focused on African American and Latino health.[9] Some medical genomicists began to seek new forms of admixture mapping technology that would distance their work from this one. Under public scrutiny, DNAPrint scientists had to explain why they were selling a technology that could give law enforcement a means to target specific races, produce race-based dragnets, and encourage other forms of racial profiling.

The developers of the technology embarked upon a campaign to distance it from racial connotations, explaining it as concordant with the publically sponsored HapMap framework. Shriver and Frudakis, in many newspapers and scientific forums, were quoted making overt statements against genetic concepts of race.[10] DNAPrint removed all mention of race from its website and business reports for several years.[11] It also provided a translation of lay terms into scientific terms, as in this publicity flyer:

> DNAWitness™ is a novel DNA test that determines the bio-geographical ancestry (BGA) of a sample. The test paints a picture of the relative amounts of European, East Asian, Native American and Sub-Saharan African ancestry of a given DNA sample. BGA results shed light on the general characteristics of a person of interest or victim. For instance, an individual with ninety-five percent European ancestry would be expected to be Caucasian and not African American or Asian.[12]

While an explanation that makes distinctions between continental designators and familiar race categories may seem absurd, it highlights the difficulty scientists faced as they tried to reconcile the newly antiracist and race-positive aims of Census race, lay categories, and a genomic taxonomy based in geographic terms like continents. As Shriver told me, test makers felt caught between their desires for the technology to be politically correct yet user-friendly. At times, he thought, "to communicate with people who are uneducated and who you're not prepared to educate at that moment, race might be the word you would need."

The media uproar led company scientists to question their own role in perpetuating racist concepts and to face their own personal responsibilities to get race right. Citing "nineteenth century stuff on black diseases" and "Negro problems," Shriver said he wondered, "That's my ancestry, intellectually, [but] I don't believe in racism, so how can it be race science?" Frudakis prepared diagrams of his own admixture results and those of his family members to share on the company website. Like many genomicists working at this turning point in race-based medicine and genomic policy, company scientists struggled to align their definition of race with critical thought and ideas from the public domain.

The solution that the genomicists at DNAPrint settled on was to publicize both a definition of biology that incorporated sociological understanding and a definition of race that considered both biological and social factors. On their website they wrote:

> Race in general usage includes both a cultural and biological feature of a person or group of people. Given the fact that physical differences between populations are often accompanied by cultural differences, it has been difficult to separate these two elements of race. Over the past few decades there has been a movement in several fields of science to oversimplify the issue declaring that race is "merely a social construct." While indeed this may often be true, depending on what aspect of variation between people one is considering, it is also true that there are biological differences between the populations of the world.[13]

This story is characteristic of the larger move in the genomic sciences toward recasting race as a multifaceted entity worth understanding from a genomic perspective. By the middle of the decade of the genome, one could say that the field had assumed a *sociogenomic* approach to race—a view that race has a genetically determined component and a socially constructed component. Scientists facing scrutiny from editors, social scientists, public health experts, and bioethicists adopted social epidemiological frames for investigating the molecular, and made racial health disparities a focus of their work.

I will show that while the genome projects were becoming ever more racialized in the first years of that decade, genomicists' engagement with the wider scientific community, with critics of genomics, and with the public encouraged them to turn race-consciousness into a sociogenomic paradigm. As scientists were called on to solve public health dilemmas around race and health disparities, they made the study of social problems, identity, representation, and minority recruitment a mainstay of their race-positive investigations.

Finally, as individual genomicists have developed sociogenomic research, they have found a voice for their own political investments. As in the example of

DNAPrint, scientists are forging ethical postures through the give-and-take of public engagement. As they respond to questions about racial sampling, inclusion, and the power of genomic research to change identities and social beliefs, genomicists draw on values that help them create ethical toeholds in their practice, communicate with other experts about race, and experiment with activism from a scientific standpoint. I describe the making of racial science then in terms of values-based pragmatism and a distinct personalizing ethos.[14]

Public Health Debates

While research and recruitment policies were changing in and across genomic institutions, in the early 2000s, a slew of debates with the broader science community reinforced genomics' turn toward social and biological aspects of health as a basis for genomic inquiry. Genomics' entrée into public debates came in 2000, as scientists were preparing their celebratory remarks about the human genome. The field's leading journal, *Nature Genetics*, released an editorial calling on scientists to reconsider whether they were reifying race as genetic and to define their sampling choices:

> From now on, *Nature Genetics* will therefore require that authors explain why they make use of particular ethnic groups or populations, and how classification was achieved. We will ask reviewers to consider these parameters when judging the merits of a manuscript—we hope that this will raise awareness and inspire more rigorous design of genetic and epidemiological studies.[15]

The editorial implied that race was a "pseudo-biological variable" that scientists should avoid in their bench-to-bedside efforts. Rather than simply change their reporting practices, some genomicists decided that it was time to use what tools genomics had to make their own definitive statement about race. They proceeded to attempt to directly measure race using new genomic technologies in order to determine its validity.

The University of Ferrara's Guido Barbujani and researchers at the Max Planck Institute examined whether continental groups, and an individual's continental origins, could be inferred from genotypes based on two different sets of markers. They found that while both sets of markers produced broad genetic clusters on a continental scale, each set produced *different* clusters. Furthermore, neither cluster had greater than 70 percent predictive value for an individual's continental origins. In other words, varying the study designs produced varying results. The researchers concluded that the genome's structure was not definitive enough to support the notion of clear, separate races.[16] In another study, David Goldstein and a team of University College London and Oxford scientists blindly grouped human genetic variation on the basis of drug response genes in order to see if genetic clusters matched lay categories.[17] These were the first-ever *genomic evaluations of social descriptors*. Goldstein and his collaborators found that "commonly used ethnic labels (such as Black, Caucasian and Asian) are insufficient and inaccurate descriptions of human genetic structure."[18] Of the study, a principal investigator of the NIH Pharmacogenetics Research Network, Howard McLeod, remarked: "It is no surprise that skin pigment is a lousy surrogate for drug metabolism status or most any aspect of human physiology."[19]

The editors of *Nature Genetics* again urged other genomicists to use "race-neutral" study designs that would capitalize on the growing body of data from the genome.[20] Editors from the *Journal of Adolescent Health*, the *Archives of Pediatrics and Adolescent Medicine*, and *Genomics* followed in tow with policy statements asking contributors to be able to justify their use of race.[21] This research and the policies that sprung from it sent the message that genomics should be exemplars but also the ultimate arbiter of race, in science and beyond. Scientists and editors agreed that genomics could serve as a corrective to conceptual errors in biomedicine and the public sphere, thus becoming the new science of race.

Though these studies seemingly demarcated the social from the biological, another contribution to the debate added an explicit sociological angle when a team of California researchers made the controversial claim that the genome proved that race was, in fact, *real*. Against prior findings, Stanford University's Neil Risch and Hua Tang, and University of California collaborators Esteban Burchard and Elad Ziv, argued that genetic variation correlated perfectly well with commonsense continental races, and that despite the admixed character of some federally recognized races, these populations did in fact constitute separate reproductive populations—"there is great validity in racial/ethnic self-categorizations, both from the research and public policy points of view."[22] They further argued that a lay notion of race was more medically interesting than genetic cluster data, because the former captured important social, cultural, behavioral, and environmental variation in addition to biological variation. They warned that ignoring race would result in an oversampling of "the Caucasian majority" and lead researchers to ignore minority health. They

closed with a call to action: identify genetic differences between races, starting with minorities, and put this to work in public health. This was the first of
several strong "pro-race" appeals to come from genomics.[23] Its message precipitated a sense within the field that genomics was responsible for understanding
racial health disparities.

Stanford University's Marc Feldman, his former students Noah Rosenberg
and Jonathan Pritchard, and other luminaries like Yale University's Kenneth
Kidd and CEPH's Howard Cann subsequently provided the data to back up
these claims. In *Science*, they reported the existence of race according to genetic clustering technology performed on Human Genome Diversity Project
samples: "We have found that predefined labels were highly informative about
membership in genetic clusters, even for intermediate populations, in which
most individuals had similar membership coefficients across clusters."[24] They
recommended that "self-reported population ancestry," or racial and ethnic
self-identification, be used clinically as a proxy for genetic ancestry to avoid
more "intrusive" measures like sampling individuals' DNA. In other words,
though self-identification was widely agreed to be subjective—reflective of an
individual's social persona and malleable throughout the life span—scientists
were willing to subsume their methodological and social needs by accepting
self-identification as proof of the unchanging genetic core of an individual.
Race was considered real enough for biomedical purposes.

With these studies, news of the genomics of race lit the pages of newspa-
pers worldwide, again widening the field of debate and inviting more scientists
to reflect on the social implications of their work. The *Toronto Star* declared:
"Race Seen as Crucial to Medical Research." The *Los Angeles Times* announced:
"Race Is Seen as Real Guide to Track Roots of Disease."[25] In a summary article,
New York Times science editor Nicholas Wade translated the *Science* diversity
maps into color-coded continental races of "Africans, Europeans and Middle
Easterners, East Asians, Melanesians, and American Indians."[26] As the media
quaked with news of races found, genomicists became regular guests in science
columns and news programs.[27]

Whether arguing for or against the use of race, genomicists were fast becoming experts on race. Risch maintained, "we need to value our diversity
rather than fear it"[28] and "[genetics] obviously correlate with the major races
as they have been intuitively identified."[29] The Diversity Project's Mary-Claire
King framed the findings as a paradox: "Everybody is the same; everybody is
different."[30] The National Cancer Institute's Stephen O'Brien agreed that popu-

lation geneticists knew all along that race was a valid genetic concept. He interpreted Risch's remarks as wresting the concept of race from the hands of well-intentioned, politically correct bystanders "that will hurt epidemiological assessment of disease in the very minorities the defenders of political correctness wish to protect."[31] Goldstein, however, countered: "you want the best representation you can find. . . . That's an argument we will have in the scientific literature and Neil will lose."[32] Aravinda Chakravarti similarly asserted: "We can't wish away these boundaries. But I'm not convinced that knowing these boundaries is necessarily useful for genetic research."[33]

By the time the National Human Genome Research Institute published the final draft of the human genome and commenced the International HapMap Project, race-positive messages abounded. The *Washington Post* marked this transition as one from a race-free project to a "roadmap for future [race-based] studies."[34] Yet, speaking on behalf of the project, David Altshuler publicly doubted that it would find a widespread existence of race-specific alleles or any actionable differences by race.[35] Planners surmised that the project was leading the field to replace race with the concept of ancestry. They made clear that the directors of this project were dually committed to inclusion and to creating better models of human variation.

As the decade wore on, research moved further out of the lab and into the domain of public health, where race was routinely explored with genetic and social epidemiology. Large-scale epidemiological studies, such as the Vax-Gen HIV vaccine trial, were gaining attention for their race-stratified results.[36] Several new studies reported a genetic basis for the disparities in cardiovascular disease and cancer disproportionately affecting African American men.[37] Scientists developed finer methods for dealing with population substructure, or undetected divergent ancestry in the cases and controls of large epidemiological datasets,[38] while moving toward studying the comparative genetics of normal traits that contribute to everyday health.[39] Now that the Genome Project was finished, scientists faced public pressure to deliver health-relevant findings. How to include minorities and how to address the health issues that mattered to them grew urgent.

In the pages of the *New York Times* and the *Wall Street Journal*, American Enterprise fellow Sally Satel ramped up the field's controversy with articles entitled "I Am a Racially Profiling Doctor" and "One Nation Under Racist Doctors? Don't Believe the Media Hype."[40] A psychiatrist, Satel accused the new science of promoting colorblind care. Describing real-world medicine as cognizant of race, she gave examples of diverse medical settings where clinicians carefully used race to determine how to treat patients. Satel concluded: "So much of medicine is a guessing game—and race sometimes provides an invaluable clue. As citizens, we can celebrate our genetic similarity as evidence of our spiritual kinship. As doctors and patients, though, we must realize that it is not in patients' best interests to deny the reality of differences."[41] Her comments depicted genomicists as removed from, thus less sensitive to, the day-to-day troubles that doctors and patients faced. She also drove home Risch and colleagues' position that researchers and medical practitioners needed to help minorities in any way they could. From this point on, genomic debates would hinge on the dilemma of racial inclusivity versus scientific acuity.[42]

Once more, controversy resounded in the *New England Journal of Medicine*. In a commentary that stressed the inconclusive state of most research on race, deputy editor Bette Phimister said, "it seems unwise to abandon the practice of recording race when we have barely begun to understand the architecture of the human genome and its implications for new strategies for the identification of gene variants that protect against, or confer susceptibility to, common diseases and modify the effects of drugs."[43] From her former post as editor of *Nature Genetics* Phimister had months before told the *New York Times*, "Risch's point that there is a high and useful degree of correlation between ethnicity/race and genetic structure, is well taken, and one with which we agree."[44] Hers was a practical message for epidemiologists, medical practitioners, and others in clinical research: use race until the genomic jury is in.

The journal featured another commentary by Risch, Tang, Burchard, Ziv, and other notable figures like Joanna Mountain, which reiterated the necessity of using race in biomedical research and clinical practice, especially given the emerging epidemiological thrust of the field. After arguing that gene frequency variation in ethnic groups was "typically not as great" as that in races,[45] they pointed to the long-established practical utility of this variation: "There are racial and ethnic differences in the causes, expression, and prevalence of various diseases. Rather than ignoring these differences, scientists should continue to use them as starting points for further research."[46] Here genomicists were staking the field's raison d'être on inquiry into racial health disparities. They maintained that genomics would provide knowledge that the entire biomedical field desired and much needed. Interestingly, contrary to their claim in *Genome Biology* that the United States possesses clearly defined continental races, they highlighted the need for recognizing American admixture. Still they noted their support for the rigid Census taxonomy, even going so far as to denounce a political initiative that would have prohibited racial classification in medicine and state business in California. Again they warned that an absence of racial signifiers would prevent monitoring of minority inclusion, which itself would lead to medical Eurocentrism. From their perspective of genomic epidemiology, race was the best variable for the job.

Giving a dissenting opinion, cardiovascular epidemiologists Richard Cooper and Jay Kaufman, and Diversity Project founder Ryk Ward, countered that race was too imprecise to be of use in the laboratory or the clinic.[47] Though they acknowledged the correspondence between continental genetic clusters and Census race, they argued: "Most population-specific microsatellite alleles are unlikely to be functional; rather, like a last name, they merely help to verify the geographic origin of a person's ancestry."[48] Doubting the validity of the BiDil study design, they further said, "if you really need to know whether a patient has a particular genotype, you will have to do the test to find out." Still, like Risch and colleagues, Cooper, Kaufman, and Ward charged the field to assume the dual responsibility of being "the beacon of science" in informational terms—by unlocking the secret of life—but also in *social* terms, by proving that science could be self-conscious and reflexive about its role in society.[49] They cautioned that "pro-race" scientists were siding with racists of the eugenicist ilk and "violat[ing] the principles that give science its unique status as a force outside the social hierarchy."[50]

The National Human Genome Center's release of their plans to conduct a large-scale epidemiological study on African Americans provided yet more support for a sociogenomic paradigm that would engage minority communities to produce their own genomic revolution. Risch professed the hope that racial minorities would follow after Ashkenazi Jews in taking control of their genomic information. "The more minorities we can get doing the research, the more these [racial] issues will dissipate." Risch's comments bespoke not only the field's intensifying interest in recruiting minority researchers to access vulnerable populations and make genomics palatable to them but also the need for genomicists to consider minority needs from minorities' own perspectives.[51]

In June 2004 the editors of *Nature Genetics* issued another editorial entreating authors to move from race to ancestry and ethnicity. In "The Unexamined 'Caucasian,'" the editors directed scientists to consider "descent, continental origins, and admixture" as ancestry, and social and cultural factors that shape "phenotype, migration, and reproductive patterns" as ethnicity.[52] To describe health inequalities as a result of racial discrimination, they asked that research-ers also consider using ancestry and ethnicity. Finally, the editors requested uniform clarity in descriptions of study populations, so that findings could be transferred to other contexts and better generalized.

Four months later, the *Journal of the American Medical Association* followed suit with a new set of "Author Instructions" requiring authors to clearly define and justify their use of race, both in terms of their study and in the context of previous literature. Deputy editor Margaret Winker stated:

> When reporting race, ethnicity, or both, authors should describe who desig-nated race and/or ethnicity for an individual; self-designation generally is preferred. Authors should indicate whether the options for designation were closed or open. If the options were closed, authors are asked to provide what the options were, whether categories were combined, and, if so, how. . . . Finally, authors should indicate why race and/or ethnicity is believed to be relevant to the particular study.[53]

Winker acknowledged that biomedical research involved a series of careful choices made in the context of work with federal mandates, self-determined communities, and changing demographics. However, she made clear that the journal required transparency on these matters. Winker concluded that "deter-mining race" was but a first step in understanding health.

Only one of these communiqués challenged the role of *genomics* in deci-phering race. Most policymakers simply requested that scientists better explain their findings, and many referred to genomic insights about race to legitimize that request. The genomic science of race was fast becoming an authoritative and exalted source of racial knowledge but also a benchmark for claims about reality beyond the halls of genomics proper.

In public, scientists discussed visions of a minority-focused translational genomics. Reflecting on the National Human Genome Center's plans for the African American biobank, Cooper again warned of the "exaggerated empha-sis" given to race. But the project found a supporter in the typically anti-race Lynn Jorde, who with Burchard called for more inclusion by such innovative means. These scientists supported the "by the people, for the people" aspect of the project but also the field's face-forward approach to health disparities and the biology of race. Implying that genomics could best get to the latent core of race, Burchard remarked, "the risk of not looking under the hood far outweighs the risk of potential misuses." These comments show that interpretations of the viability and efficacy of new translational interventions were deeply entwined with values about redefining race and sending a message of racial equality.

As the public was becoming familiar with genomic proclamations on race, newswires began buzzing with reports of the BiDil trial's hasty close. The October 29, 2004, *New York Times* cover story exclaimed: "We can treat your heart disease . . . if you're black."[54] The *Wall Street Journal* called it "reality-check time" for scientists who believed that race was biologically meaningless.[55] These articles and others featured interviews with the Howard conference participants, including Francis Collins, who said the connection between race and genetics was too "blurry" to make it medically useful, and Jorde, who said: "We find African genes in Scandinavians, and English genes in Chinese. There is no such thing as a genetically 'pure' population." Goldstein cautioned against targeting race in place of genes,[56] while Charles Rotimi worried that there was no definitive research showing that BiDil didn't work in "whites" too. In the *Lancet* he quipped, "Who is African-American?" In a front-page op-ed for *USA Today*, Cooper asserted, "The last thing we need is greater acceptance of race in the name of 'science' when there is no scientific evidence."

Yet as genomicists began facing the reality of doing genomics in a health disparities framework, many retreated from a staunch position. For example, Altshuler was quoted as supporting the use of race as a "last resort," while Goldstein approved using race as "an interim measure."[57] Cohn and the BiDil investigators were claiming that direct causes were not necessary to produce sound medicine.[58] Though genomicists disagreed, they increasingly faced supporters of "meantime proxies" in their new partnerships with minority advocacy groups. They also remained squarely in the limelight, often expanding their public role into policymaking and advocacy.

If the state of pharmacogenomics and biotech was any indication, genomics was indeed heading for a deeper reliance on race to solve health disparities and the puzzle of gene-environment relations. The private companies that served as collaborators on major national and international projects were structuring their studies by race, and pharmacogenomics was looking at health behaviors associated with race. While basic research might oppose using shortcuts to variation, clinicians valued them. As William Evans, director and CEO of St. Jude's Research Hospital, and Mary Rellings, chair of pharmaceutical sciences there, commented in *Nature*:

> The full elucidation of genotype-phenotype associations may eventually lead to the use of genetic markers alone and obviate the use of race to adjust for the po-

tential confounding of underlying population substructure. But self-identified race information may be associated not only with genetic differences, but also with non-genetic influences on drug effects (for example, diet and salt intake), in which case adjusting for race or ethnicity may have a use that goes beyond genetic implications.[59]

The landscape of the field was itself changing: one of the most central academic research institutes, the Whitehead, emerged as a privately endowed organization, the Broad. The distinction between government, academy, and industry was fast becoming a relic of the past.

In the weeks surrounding the publication of the BiDil study, the *New York Times* released three articles on the drug's success and the advent of race-based medicine.[60] Though the articles explained the nature of the trial and showcased the views of the drug's makers, the majority of coverage was devoted to battles between genomicists over race, ancestry, disparities, and discrimination. One queried Craig Venter and Collins on their millennial White House claim that the genome was race-blind. Venter maintained a strictly anti-race position: "I don't see that there's any fundamental need to classify people by race. What's the goal of that, other than discrimination?" Collins, on the other hand, replied that he wanted genomics to take on more investigation of race and its relevance to health disparities and medicine: "We need to try to understand what there is about genetic variation that is associated with disease risk, and how that correlates, in some very imperfect way, with self-identified race, and how we can use that correlation to reduce the risk of people getting sick."[61] Collins made clear that the field was ready to begin investigation into the gene-environment relationship between race and health, and to take the reins of an inquiry that thus far had come up null. He further developed the pervasive contention that genomics was the *only* field that could be trusted to solve society's race problem:

> I think our best protection against [racist science]—because this work is going to be done by somebody—is to have it done by the best and brightest and hopefully most well attuned to the risk of abuse. That's why I think this has to be a mainstream activity of genomics, and not something we avoid and then watch burst out somewhere from some sort of goofy fringe.[62]

Quoting from the Bible, Collins depicted the Genome Research Institute's leadership in racial issues as a crusade: "And you shall know the truth, and the truth shall set you free." In agreement, the article concluded: "Rather than abort a whole field of research because it might bolster cranks and demagogues, maybe one solution to our national angst over race is to let scientists hunt down the facts—facts that will no doubt affirm, one way or another, that the human genome is indeed our common thread."[63] At mid-decade, the transformation was complete. Scientists were embracing the task of investigating race from a social and biological perspective and were certain that genomics should be the field to do so.

Critical Engagement

The integration of race into ELSI's funding programs has also been an important means by which genomic racial expertise has been formulated into a sociogenomic research framework. Early on, it was primarily through ELSI programs that social scientific and bioethical problematics were incorporated into the planning processes of the major human genome projects, and thus brought to the attention of genomic elites. For example, sociologist Troy Duster was a prominent collaborator in the 1990s' planning meetings of the Human Genome Project thanks to his involvement in the first rounds of ELSI-funded social research and his rising influence in the social sciences and medical humanities. In a well-known instance, Duster scandalized planners of the Human Genome Project by notifying them of a little-known forensic report on the genomics of race. The report suggested that criminologists needed only three segments of DNA to determine whether a suspect was "Caucasian," "Afro-Caribbean," or "Asian Indian."[64] Collins, then the director of the Genome Project, later called the moment "chilling." Until then, he and the other scientists had no idea that genomic research was being put to such use. Duster began calling on genomic researchers to consider their own hand in perpetuating racial beliefs.[65] As mentioned in Chapter 2, this first round of ELSI research investigated how various racial groups experienced genetic testing, screening, and counseling; it did not examine how racial ideas would be affected by genomic research. Behind closed doors, though, Genome Project planners were made aware of the potential for the reification of race on the basis of genetics.

At the millennium's turn, as genomics was fast institutionalizing its own racial policies and developing an interest in health disparities research, two ELSI research projects yielded findings that suggested a porous boundary between identity and genes, which were thought to be in discrete social and biological domains. Bioethicists Carl Elliott and Josephine Johnston's research into ethnicity, citizenship, family, and identity in the wake of the Human Genome Project found that recreational ancestry estimation was threatening to usurp cultural and historical ways of determining indigenous American identity.[66] The conference they hosted in June 2002, at the University of Minnesota, led to a special issue of *Developing World Bioethics* in which participants from genomics, social science, and the humanities presented their views on the essentialist threat posed by reconstructions of African ancestry. Elliott and Johnston urged scientists and the public to consider the potential impact of ancestry estimation on individual and group identity "before, during, and after such tests are carried out."[67] In the United Kingdom, social anthropologist George Ellison and sociologist Ian Reese Jones also contributed a study on racial genetics and identity. It looked at how scientists were using subjects' self-identified racial and ethnic concepts. Their title search of three biomedical journals and sixteen influential human genetics journals found that in biomedical journals, articles focusing on genetic determinants were more likely to use race than those examining nongenetic determinants, and that across human genetics journals the use of lay concepts of race and ethnicity was increasing.[68] They recommended that social identity, with its "limited internal and external validity," not be used as a proxy for genetic variation.[69] Both these research teams beseeched genomicists to consider the fluidity of identity and the impact genomic science would have on it.

At the same time, ELSI-sponsored communications analysts showed genomicists that lay audiences brought their own prior understandings, values, and political beliefs to bear on genomic information, fusing biological and social expertise. In a series of NIH and CDC-sponsored studies on lay perceptions of genetics and race, Celeste Michele Condit and several teams of researchers found that lay audiences find ways to process genomic knowledge and turn it into personal know-how.[70] Seeing how certain focus-group members reacted to genomics news reports, Condit and colleagues warned genomicists that patients' knowledge of ancestry and suspicion of race-designated drugs would likely impact the emerging pharmacogenomics industry just as genomic research would shape how lay audiences would treat their own health and view the biology of others.[71] Again scientists were entreated to think about their findings as entwined with lay processes of meaning-making.

Early in the decade, health disparities experts from an array of social epidemiological, social scientific, philosophical, and policymaking fields asserted that genomics should eschew race completely.[72] Critics argued that after the Genome Project any persisting belief in race had to be residual genetic determinism or, worse, scientific racism.[73] They also focused on the rhetorical power of genomics,[74] and argued for race to be used only to document the persistence of racial discrimination. Some worried about trickle-down effects in the clinic, such as racial profiling.[75]

Tracing citation trails among genetically deterministic articles, legal scholar Jacqueline Stevens cautioned that genetic hype overpowered disclaimers about the social nature of race. She proposed

> that the National Institutes for Health issue a regulation prohibiting its staff or grantees from publishing in any form—including internal documents and citations to other studies—claims about genetics associated with variables of race, ethnicity, nationality, or any other category of population observed or imagined as heritable, unless statistically significant disparities between groups exist and description of these will yield clear benefits for public health, as deemed by a standing committee to which these claims must be submitted and authorized prior to their circulation in any form beyond the committee.[76]

In other words, bar science writ large from using racial terms unless "a standing committee" of health disparities experts could agree that the research merited a racial taxonomy. This suggestion posited nonscientists like social scientists and bioethicists as the experts on race and put them in a position to police the boundary between social and biological factors pertinent to race.

Critics like Stevens based their stance on the view that the "population" in population genomics was "new wine in old bottles."[77] As Jenny Reardon put it: "Scientists never did give up on the biological meaning of race. Instead, they sought to reconstruct its meaning and definition in a manner that would enable them to use it to advance the stores of human knowledge while distancing themselves from what they perceived as the ideological agendas of powerful social interests."[78] Philosopher Lisa Gannett similarly questioned "the prevailing historical understanding that scientific racism 're-treated' in the 1950s."[79] These critics presented genomicists with second-order observations about their behind-the-scenes dialogues and policy disputes, arguing that contemporary scientific racialism was a legacy of past scientific racism.[80] Their findings seemed to necessitate an extrication of race from all biological research.

Yet the push for scientists to take up health disparities and minority health research also created a counterpressure of encouragement for a race-positive, sociogenomic approach. Instead of asking scientists to remove any consideration of race from their research, critics have since embarked on a series of dialogues that will generate better ways to approach the social construction of race in the body.[81] Biologists Anne Fausto-Sterling and Lundy Braun have offered an alternative, systems biology understanding of health with which scientists can better account for the biological feedback loops of racial discrimination. Fausto-Sterling has said, "it is not that *different* biological processes underlie disease formation in different races, but that different life experience activates physiological processes common to all, but less provoked in some."[82] Braun has similarly explained:

> Genes are only one component of the complex biology of an organism, and the regulation and expression of genes are subject to extensive modulation—by other genes as well as by non-genetic internal and external factors. To fail to appreciate this view of biology is to miss important insights into the physiological functioning of the human organism and its relationship to the environment.[83]

Though critics like Fausto-Sterling and Braun intend to move the focus of racial research away from genomics, scientists have interpreted this as a call for a more proactive science of race. With their social science and bioethics interlocutors, genomicists have begun remaking race into a new object. Part social, part genomic, a sociogenomics of race is being created.

The Science of a New Race

At the heart of the sociogenomic paradigm is the notion that genomic ancestry explains the feedback loop created between social ideas of race and individual health. As Duster put it, "race, used as a stratifying practice, has produced what could be read as 'negative' health practices and 'biological outcomes that make it impossible to completely disentangle the biological from the social.'"[84] Scientists maintain that populations do not have to be discrete or consistent to be of interest to researchers. Morphology is meaningless. Populations overlap. The basis for membership does not need to be clearly defined. For instance, scientists can map ancestral groups by genetic clusters just as they are proving that clinal or graded variation is the organizing principle of human variation.[85] Flexibility and plasticity are built into the sociogenomic model.[86]

This framework follows the logic of an argument for the genetic reality of race, and its retention in genomics, posed by philosopher of science Robin Andreasen in 2000. Andreasen claimed that a cladistic, or evolutionary tree–based, notion of race was *compatible* with the social constructionist version of race: "[I]f popular conceptions prove to be biologically unjustified, objective biological races might still exist. The objectivity of a kind, biological or otherwise, is not called into question by the fact that ordinary people have mistaken beliefs about the nature of that kind."[87] Andreasen suggested that race was actually a logical representation of phylogenetic, or evolutionarily real, populations. She explained that both "common sense" notions of race *and* biological taxonomies based on phenotypic qualities were inaccurate because there were no "nonarbitrary" qualities to base groups on, but that a taxonomy based on historical genetic relations of ancestry could be less arbitrary. Andreasen maintained that even if a given cladistic system turned out to be flawed, it didn't mean there were no historically true relationships of ancestry out there to be found. Though her vision of race did not anticipate the complexity of the sociogenomic paradigm, her argument that social constructivism does not deny the existence of biologically real kinds has become a pillar for future positions.

A new race demands a new science; thus, concerned scholars of all backgrounds are sharing strategies to formulate a critical study of race. As Duster has cogently summed, "you can't be creating ethnically based medicines, which is what a lot of biomedical research is about, without also doing some sociology."[88] To this end, he has recommended that scientists avoid thinking in binary terms:

> By heading toward an unnecessarily binary, socially constructed fork in the road, by forcing ourselves to think that we must . . . choose between either "race as biological" (now out of favor) and "race as merely a social construction" we fall into an avoidable trap. A refurbished and updated insight from W. I. Thomas can help us. It is not an either/or proposition. Under some conditions, we need to conduct systematic investigation, guided by a body of theory, into the role of "race" (or ethnicity, or religion) as an organizing force in social relations, and as a stratifying practice. Under other conditions, we will need to conduct systematic investigation, guided by a body of theory, into the role of the interaction of "race" (or ethnicity, or religion) however flawed as a biologically discrete and coherent taxonomic system, with feedback loops into the biological functioning of the human body; or with medical practice. The latter studies might include examination of the systematic administration of higher doses of x-rays to African Americans; the creation of genetic tests with high rates of sensitivity to some ethnic and racial groups, but low sensitivity to others; and the systematic treatment, or lack of it, with diagnostic and therapeutic interventions to "racialized" heart and cancer patients.[89]

Duster's strategy allows for a flexible antiracist racialism. It suggests that bio- logical and sociological methods should go hand in hand in all analyses of race.

From his seat at the NIH, Rotimi has been a critical figure in the implemen- tation of social research in genomics. Explaining the gene-environment modus operandi in race, he has said:

> People who socialize together tend to eat things that are similar. . . . Even our culturation in terms of who goes out to do a 10K versus someone who stays in front of the TV, or if you notice, someone who sits around your kitchen and tell[s] stories. . . . Those have been passed on [through] health [habits]. . . .
>
> So if you want to look at a disparity, you have to look at income. You have to look at education. You have to look at why is it that you find certain people more in jail than others. Why is it that certain people are in a certain business more than others? Why is it that certain people tend to live closer to dumps than others? You have to look at the social structures.

Rotimi comes from an epidemiological background; however, even statisti- cal scientists and classically trained population geneticists are moving rapidly toward the gene-environment model. Some are researching environmental factors, like exposure to toxins. Others are adding traditional social epidemio- logical variables, like nutritional statistics, into their study designs. More re- cent developments include doing institutionally in-house social and behavioral research.

Comments from Rick Kittles, who does not have an epidemiological back- ground but is trying to decipher the relationship between skin color, oncogenes, and diet, demonstrate that such scientists are deeply interested in social epidemiology:

> I firmly believe that the bulk of health disparities are due to social determinants. Now, many say that if you study blacks with prostate cancer you are assuming that there is a strong biological determinant. No, I am assuming that there are some risk factors in that population that make them at high risk, and I want to tease out what are those risk factors. Maybe it's eating burnt meat. Maybe it's barbecue, charcoal, or whatever. But if I don't study them, I won't know that.
>
> Or if I say that it's not biological, "let's not study blacks," that's doing them a disservice. . . .
>
> I'm becoming more and more aware of the social determinants. And plac- ing that, now, into the context of these designs is important to me.

In studies like the 2010 Association of Plasma Fatty Acids with Prostate Cancer Risk in African Americans and Africans,[90] Kittles attempts to measure what he calls "the biology of race" by accounting for discrimination and variable health behaviors—aspects of health he feels are a primary issue for African Americans. In this context, the term "black" is considered to be useful.

As Mark Daly, the Pfizer Fellow in Computational Biology and creator of the influential GENEHUNTER program at the Broad Institute, explained: "It is often difficult to distinguish whether one's genetic continent of origin is the risk factor or if it's simply the access to healthcare or how seriously your medical issues get taken by the medical establishment, depending on what your background is. But there are some reasonably well-documented instances where it's likely to be genetic differences." Drawing from his expert knowledge of family genetics, he gave the examples of continental variations in prostate cancer and autoimmune disorders. Here Daly teases apart the genetic and the social, acknowledging the racial systems of inequality that contribute to health disparities, such as differential access and treatment. Daly's remarks convey the balanced view that scientists desire to take between nature and nurture. They also confirm that genomicists widely assume that the genetically significant portion of health does hold a continental valence.

Daly discussed incorporating a panoply of social epidemiological measures into genomic models, with race being but one: "There are so many elements of epidemiologic data collection—collecting information about people's diets, upbringing, and so forth—that need greater attention, maybe greater technological advances so that we could bring some of those in as well." He and many others imply that genomic computational tools are so robust that they will soon take over the social epidemiological world of disparities measurement. Though scientists agree that some variables are suited to the genetic portion of research while others are better kept for sociological concerns, it is unclear how those variables will be separated in a hybrid epidemiological study.

Genetic epidemiology, a science formerly reserved for hereditary diseases like Huntington's disease and cystic fibrosis, is now being called to bridge the biological and social gap in dealing with racial health disparities. The primacy of these methods can be witnessed in the launch of research centers dedicated to capturing social and genomic data about minority health. Harvard and Massachusetts General Hospital's Center for Genomics, Disparities and Vulnerable Populations; Howard's National Human Genome Center; the University of California, San Francisco's Center for Pharmacogenomics; and the National Human

Genome Research Institute's Center for Research on Genomics and Global Health all conduct in-house social epidemiological research directed by genomicists. These centers make up a network dedicated to using gene-environment models, where race is placed at the center of genomic inquiry. Some examples of their focal projects are the Genetic Epidemiology of Breast Cancer Among African American Women, the Genetics of Podoconiosis, and Tailoring Smoking Cessation Treatment by Genotype: Implications for Ethics and Clinical Practice.

The appropriation of social epidemiological work quickens the transfer of activist strategies across the science and lay divide. In discussing the burden of the history of scientific racism, for example, scientists have described using race in strategic essentialist terms akin to the strategic essentialism that marked postcolonial antiracist humanism in the late 1990s. Chakravarti has said:

> There is race, there's absolutely no doubt. There's race, there's race-typical thinking, there's racism! All of these things exist and I am very sure of that— they exist in the social sense. There is nothing to say that science cannot show that there are similarities and differences within and between those races, but that's not the . . . only thing to study with respect to genetic differences of human beings.

Scientists like Chakravarti acknowledge the social reality of race and the inadequacy of race in the same breath. Still they adhere to the epidemiological utility of race and its manipulation in recruitment, because they see the sociogenomics of race as alarmingly understudied.[91] Evincing the urgency with which scientists view sociogenomic research, Feldman has offered: "There are clear differences between people of different continental ancestries. It's not there yet for things like IQ, but I can see it coming. And it has the potential to spark a new era of racism if we do not start explaining it better."[92] Like Collins above, Feldman and Chakravarti believe that it is the *genomicist* who must tackle the issue of racial difference.

Following on their commitments to minority health, nearly all the scientists who spoke out against propagating a genetic concept of race—what I will call "race-averse" elites—attested to using a racial framework in their clinical or epidemiological work. In fact it was precisely these race-averse scientists who gave the most elaborate impressions of the social reality of race—referring to racism in society, discrimination based on stereotypes, even internalization of stereotypes and social organizing around the idea of race. They described a common belief that racism creates a biological feedback loop in the body:

a racialization of biology due to social factors of "discrimination, racism, and stressors." Georgia Dunston remarked:

> We have not fully described how one's perception of stress also influences the physiology of the system that feels threatened, okay? And how the body wasn't designed to live every day with the sense of threat! And how, when you constantly (as we do in chronic diseases) maintain a kind of view of yourself in your context, eventually how that shifts the whole balance of response that puts you in a different physiological state to what is going on.... What is the internal influence ... for a person who, because of certain biological characteristics, perceives themself or their continuity or integrity as being threatened—whether it is getting a job, losing a job, getting an insurance, losing insurance?

These scientists attempt to use race as a *dependent* variable—something to be explained by other measures—just as much as an independent variable or causal factor of health. For example, Dunston's work on the Genomic Research in African American Pedigrees Project was designed to explore the common health issues that African Americans face by creating a representative reference database of African American DNA. Her lab's current gene-environment study of asthma seeks to understand how genetic variance and racism-based social variance intersect.

Genomics has become society's new science of race by taking on the mantles of diversity and disparity and learning how to manage population taxonomies in a language of social justice. Critics and scientists now agree: there is a correlation between genomic ancestry and socially constructed race. Inquiry into the malleability of identity and phenotype does not challenge the core facticity of ancestral relationships or the ability to group based on ancestry. Meanwhile, refusing to recognize the *biological* processes associated with race is seen as tantamount to scientific racism. Contextualizing feedback loop theory in an anti-colorblind politics, critics and scientists have made the view that "race matters" the new orthodoxy.

Yet these routine developments continue to worry social scientists and humanists who argue that the "socio" in sociogenomics is getting short shrift. As Jonathan Michael Kaplan has argued, genomicists only consider social factors in what otherwise proceeds as a pursuit of genetic causes, because they misunderstand the relationship of causality between social and biological factors:

> Rather than standing as proxies for ancestry or population-level genetic differences, folk racial categories should themselves be recognized as key causal com-

ponents of biological differences. That these categories are social—and that membership in a particular category is contingent upon the particular social/cultural milieu in which one finds oneself—does not imply that the categories are biologically meaningless. It does imply that there can be no straightforward causal pathway from biological facts to membership in a particular folk racial population—biology doesn't make "race." But "race" might very well make biology—the causal arrow can point from a social category to the creation of biological differences between the populations identified. And indeed, the evidence is overwhelming that, at least in the U.S. context, race membership does just that.[93]

Critics like Kaplan still find that sociogenomic, race-positive science is necessary. However, they approach the "socio" of sociogenomics in terms of *racism*, not some sequela to innate DNA processes.

ELSI teams, social science researchers, and bioethics panels operating under the auspices of the Wellcome Trust and the American Society for Human Genetics continue to fight for research into the upstream social conditions that foster racialization—issues like ghettoization, toxic living conditions and lack of access to safe housing and proper healthcare, job discrimination that leads to particular groups concentrating in low-wage and unsafe work sectors, and stress as a response to racial discrimination.[94] Meanwhile, public health researchers are exposing how racial groups show 180-degree turns in historical prevalence for illnesses like cancer and cardiovascular disease.[95] The idea is to draw attention, and research funding, to the intergenerational legacies of racial biology and to incorporate genomics as it corresponds to conditions influenced by racism. Critics like Fausto-Sterling and Duster also increasingly champion an epigenomic model,[96] or one that explores "stably heritable" traits that result from chromosomal interactions with the environment,[97] to explain the way racial discrimination is written into the body through social mechanisms. They push for research models that capture the dynamic relationship between the body and its social context.

Personalizing Medicine

The result of all these public engagements is a kind of ethical interpellation: genomicists are continually called to consider how racialization processes, personal experiences, and self-identity relate to their research. Conversations with the wider scientific community and their critics lead scientists to revisit their own store of thoughts, feelings, and memories, and their own personal stakes. Meanwhile, personal political interests rise to the surface.

Indeed, one of the most striking features of today's science of race is the way that scientists openly cope with the politics of their work by relating it to their own life events. The scientists with whom I spoke personalize the tasks of reconceptualizing race and changing scientific protocols to meet political ends. This became most apparent to me when speaking with scientists about their reasons for doing genomics. Whether talking about past events or present choices, scientists slipped into stories about their own experiences with struggles for equality. They marshaled racial memories chosen from many formative experiences to interpret their present situations and to create a coherent self-image.[98] The recall of prior experiences became the basis for interpreting genomics in an openly subjective way and emphasizing sociopolitical forces affecting the life course, health, and biology. Personal racial turmoil or acknowledgment of another's was a recurrent, nearly ubiquitous theme.

Sarah Tishkoff is one of the world's most esteemed experts on African populations. Between sampling expeditions in sub-Saharan Africa and her lab work at the University of Pennsylvania, she consults with the NIH and the National Human Genome Research Institute on their Sample Repository at Coriell, serves as associate editor for *Molecular Biology and Evolution*, and advises Henry Louis Gates, Jr. on the New Genetics and the Trans-Atlantic Slave Database at Harvard University. She has been featured as one of the "Ten Most Brilliant" young scientists in the United States by *Popular Science* magazine. In a story about her entry into African population genetics, she told me:

> I think there was a sense that [Pygmies] were representative of African diversity. So, everybody would just use these Pygmies as almost like an out-group in these studies of non-Africans. I felt like there was a *huge* deficit (and there is still) of samples from Africa. So, I was going to go making my—I don't know. That was my *passion*. To go to Africa, collect these samples, and start filling in some of the gaps!

A white scientist, and one of the few female scientists leading the field, Tishkoff identifies broadly with racial legacies in America and sees educated Americans as having a responsibility to redress racial ignorance. Her drive comes from the feeling that she can use her scientific role as a corrective for past scientific racism. She wants to move the field beyond the assumption that a single population's DNA can stand in for an entire continent, and beyond the complacency of biomedicine, which has generally failed to use African genetics in behalf of African health in its own right. Like many others, she uses the term "passion" to describe her enthusiasm and sense of responsibility for addressing lacunas in our knowledge. For genomic elites like Tishkoff, the lab becomes a site to correct the errors of the past.

A similar passionate commitment, yet one personalized at the level of social identity, can be seen in David Goldstein. In his introduction to *Jacob's Legacy: A Genetic View of Jewish History*, Goldstein wrote:

> I suspect that had I had no Jewish heritage, this work would likely have never led me into Jewish genetic history. Events, however, primed me to look for ways to translate my professional activity into some kind of connection with my own ethnic background. I remember, for example, feeling agitation and anger during the first Gulf War when Israel was being bombed by Iraq and America was instructing Israelis to sit tight. My first, and I suppose somewhat childish, impulse was to enlist in the Israel Defense Forces.[99]

Goldstein, too, sees his lab as an alternate political battleground. Yet for him there is an even tighter connection between the object of his research and his social identity. Goldstein runs a next-generation sequencing lab at Duke University that explores the pharmacogenomics of cardiovascular disease, neurology, and HIV, but he is publicly known for his research establishing the existence of Jewish priestly lineages. He has prominently figured in the major news media in recent years, especially in a series of articles in the *New York Times*.[100] Like Tishkoff, Goldstein recalls critical life events and career choices through a starkly political lens. For these scientists, it is not shameful to link one's political views, and even one's own identity issues, to one's work.

Race poses stark ethical dilemmas for science. Like all the scientists I interviewed, these scientists think through such dilemmas according to their social and political beliefs. They move seamlessly between social considerations of race and genetic explanations of population diversity, and they shuttle between political notions of race and strictly biological explanations. For genomicists, conceptualizing race involves considering different aspects of heredity, ancestry, political history, and social disparity and then creating assemblages of meaning that can respond to an array of social histories and scientific needs.

We should approach these scientists as ethical subjects responding to a fraught political context. As Kelly Moore argues in her study of postwar science's political activism, "To see scientists only as experts, however, neither captures the variety of roles they have played in politics nor enables us to understand how scientists conceive of their own actions."[101] Scientists have been "reorganizing the moral economy of science" over half a century of antiwar, feminist, and antiracist activism.[103] Challenging their peers to practice socially responsible science, scientists have themselves increasingly worked to repaint science as multifarious, politically conditioned, and even fallible.[103]

Indeed, the scientists here have come of age—and have been trained to examine human variation—at a time when racial justice has been a prominent political issue. Their way of describing their experiences counters the assumption that scientists are "people who are 'not like us.'"[104] Scientists too are cognizant of inequality and committed to racial justice. Just as with lay movements, scientists' informal coalitions can become prominent activist movements when their values link with dominant trends in social justice.[105] In fact, genomics is in step with the racial zeitgeist of what Michael Omi and Howard Winant call America's "racial democracy project"—the post–civil rights historical formation dedicated to producing law-based racial equality.[106] Working through their own normative desires for the future of race relations, which has concerned many of them their whole lives, is part and parcel of scientists' research experience. What is key here is not so much the veracity of scientists' experiential claims but rather the linkages between scientists' perceptions of their social responsibilities, their strategies for engaging the moral order, and the social context that they themselves are changing.[107] They view understanding race as a scientific challenge; in other words, they view their science as an opportunity to work through the acute problem of race. Thus, race-positive values are infused in their every project.

At the same time, the field of genomics is itself responding to prior failures in biology to produce a just and socially acceptable definition of race and human variation. At the turn of the century, when the notion that "race is a social construction" was being touted by famous geneticists like Richard Lewontin and Stephen Jay Gould,[108] critics attacked Diversity Project planners like Feldman and Luca Cavalli-Sforza for infusing the new international genome projects with racist taxonomical ideas.[109] Then, as the colorblind paradigm for dealing with issues of race collapsed amid multinational, public criticism, a nascent genomics struggled to make better sense of the relationship between race and genetic variation. Today's elite scientists attempt to build a virtuous reputation with the public and a richer public role for the field through their new science of race.

Since genomicists find themselves between the tasks of producing a definitive human taxonomy and a just racial social order, scientists oscillate between social, biological,[110] and genetically deterministic rationales for race.[111] They combine objective rationales with subjective, highly personalized ones and look for ways to square their values with the technical rigors of their practice.[112] This creates a composite of seemingly contradictory norms that frame genomic research. On the one hand, scientists remain committed to practicing in a disinterested fashion, according to shared and impersonal rules of scrutiny, and toward a common end of discovery.[113] As Frudakis insisted, though separating "your human self" from the "objectivity of your work [is] probably impossible to do," one must try. He implored, "How can one objectively ask questions about the universe when one is an advocate for a certain outcome?" Frudakis worries that in failing to suspend one's values and adhere to the shared standards of practice, scientists will produce "a crappy result." On the other hand, scientists consciously deploy ethical interests and allow personal commitments to guide them toward specific ends in certain moments of the research process, especially during formation of the research question and research design. As Chakravarti argued,

> I don't even know whether first and foremost I am a scientist; I have many other roles. I am somebody's father, I am somebody's husband, I am somebody's brother, and I have a role in society—and I am very interested in that role in society. . . . I cannot believe in things that my science says is not true, but the reverse is also the case. Just because I could do a piece of science, if I think it's socially—really, personally—not acceptable to me, I would not do it.

This prioritization of one's family role was remarked on by many scientists, who pointed to the pits other scientists have fallen into with purportedly "value-free" work on volatile topics like nuclear energy and the environment. Genomicists carefully calculate how their work will be perceived, not only by the public but by their loved ones and kin.

What is conditioning genomics foremost is a highly charged personal enthusiasm for human variation research, which bridges the realms of dispassion and interest. Scientists not only openly exhibit joy about being a part of the field's emergence at this historical moment—what many describe as the brink of racial enlightenment; they also expect others in the field to do the same. Chakravarti continued:

> I have an aim that by the time I retire from science, almost everything that I learned in my early part of my career will have been proved either irrelevant or incorrect. Genetics is clearly capable of doing that. I don't think we can mature unless we change our view and prove ourselves, our past, incorrect. That's, to

me, progress. Not "learning new things"—[but] learning new things *and proving that previous things were wrong.*

Scientists like Chakravarti described a litany of ideas and personal beliefs that they have upended with their science. Many spent time recounting *others'* personal motivations, stories, and research trajectories, even those of scientists they had not directly worked with or agreed with on matters of debate. In a critique of the marketing of direct-to-consumer ancestry tests, for instance, Rotimi wrote of Kittles:

> Kittles indicated that the need to know his ancestry and the strong desire to begin to see, feel, and live his history beyond slavery is a longing that has significantly shaped his personal drive and professional development. Becoming a biologist with training in evolutionary biology and genetics for ancestry tracing is both a personal and a professional commitment. . . . The feeling is that no matter how limited this knowledge may be, it holds the potential of helping African Americans uncover their ancestral history before the onset of slavery.[114]

In recounting each member's personal connection to the field, these scientists create a social bonding among themselves and engender a supportive, "interested" infrastructure for the field.

The genomicists I spoke with also demonstrate a willingness to excuse the limitations of the science on what they see as the path to racial truth. They emphasize that scientific pragmatism and personal values are integrated in their research. This draws together collectivist and skepticist norms, yet couched in an ethos of a commitment to racial justice. In an optimistic tone, these scientists express a do-your-best sentiment: since you don't know how racial concepts and relations will change over time, in the present "you do your best to try to be the right kind [of scientist]," as David Altshuler argues. Most speak confidently about the field's ability to overcome present shortcomings and eventually produce a correct, universal human taxonomy that will dispel fictitious racial beliefs.

Linked to the personal enthusiasm and do-your-best pragmatism is an overarching optimism. For example, no matter their views on race, scientists were optimistic in discussing the development of drugs for minorities. Goldstein expressed this best in comments about his deep belief in genetic equality:

> You can't be ignorant of the history, that science has been used as part of racist agendas in the past—you have to be conscious about all of that. But I also think that scientists whose hearts are apparently in the right place are directly

contributing to racism by acting like what's going to happen is that if we find a whole bunch of gene variants that are associated with [better] cognition, "well, we all know that they are going to be more common in white people." I don't believe that, and so I'm not worried.

Scientists like Goldstein expressed a deep faith that life's basic processes are the same for all human populations. They thus offer genomics as a corrective to prior scientific racism and the limits of social epidemiological health disparities research, which purposefully ignores genetic causal factors over fear of what will be found. Goldstein himself works on epilepsy and schizophrenia, two diseases that exhibit racial disparities and are highly heritable. As with all genomic behavioral research, findings have implications for neurological processes beyond the disease in question.

Sentiments like these also show that ethical judgments and reflexive analysis are intrinsic to genomic interpretations of research findings. Though constructing human taxonomies is not in the official job description for someone working in genomics, distinguishing human populations is. Every study determines population groupings that have the potential to look like the racist taxonomies of the past. The scientists heading up the field are aware of the power that genomics holds in influencing public views on racist taxonomies, and they want to produce a genetic science that is politically sensitive. In seeking to avoid reifying racist notions, genomicists develop a critical mind-set and an array of interpretations, strategies, and values, which they enact in the many social arenas in which they take part. They manifest a new science of race that draws on the virtues of politically engaged science forged by geneticists like Theodosius Dobzhansky and Walter Bodmer. Yet going beyond participation in political debates, genomicists also calibrate their research practices to their values and beliefs about race. In doing so, they create new avenues of social action based in an ethical scientific rationality.

The genomic style of thought can be summarized by inverting the famous feminist slogan "the personal is political" to read "the political is personal." Because politics was formerly assumed to be separate from personal life, including the informal social domains where women often worked, the slogan galvanized Americans to wake up to the political dimensions of personal life. Yet for scientists there are also personal dimensions to the political life of science that continue to remain in the dark. The science activism described here forces us to examine the "personal passions" that motivate scientists and urge them to their research; and to observe the thought processes that permit scientists to cultivate their own subjectivities through their scientific work. As scientists fashion themselves as activists, advocates, and policymakers intent on creating a climate of social justice (many seeing themselves in the midst of ethnocentric and racist social arenas), they collectively expand the bounds of scientific duty to encompass social health concerns like minority inclusion, representation, and health equality.

How does this pattern of personalization become an ethos for the field? To begin with, elite genomicists are not mere bench scientists; they are public representatives of the field and institutions they belong to. Public engagements are routine, and thus the job provides many opportunities to present one's ethical commitments in speeches, television interviews, filmmaking, and so on. Nearly all the scientists I spoke with embrace this role and insist that genomicists should be more visible on social issues and matters of race. Joel Hirschhorn, a Harvard Medical School professor, research director at the Broad Institute and Children's Hospital Boston, and a pioneer of genome-wide association, refers to it as the field's "obligation . . . for clarifying . . . and lessen[ing] societal discrimination" in light of past abuses in the application of genetics and biology. Hirschhorn engages in curricular development toward this end, and his home institutions have held numerous discussion forums on the perils of scientific racism. His words mark the connections scientists make between the knowledge-seeking moment of research and subsequent implementation of that knowledge toward specific political ends. They also convey the interest scientists take in remolding the social arena with their tools. This necessarily involves using their own scientific constructions to challenge categories that they inherit from the wider social arena.

The Many Facets of Race

By adopting the concerns of their critics and measuring their own practices against those standards, scientists have responded to mounting anxieties that genomics will reify race in biological terms. They increasingly see themselves as frontline members of a larger network in translational medicine. Studying the social characteristics of health is a basic interest of genomic research. The result is an ethos of personalization and a new sociogenomic research paradigm.

Scientists are partaking in a new science of race, because in this day and age it is socially impossible not to do so. The NIH funding streams have combined into a dominant research paradigm at the same time that scientists have been hailed into debates that provoke constant reflection on their own racial experiences. Genomicists now take the social responsibility of their work very seriously and attempt to foreground their values about race and social justice. They flexibly respond to the social context with a values-based pragmatism. This permits them to remain reflexive about their role in society as scientists and their own political agendas.

Central to the genomic ethos is an open politicization of personal experience and the ways scientific practices connect with the future that scientists want to create. The field does not adhere to the typical reputation of scientific disinterest. Rather, scientists seek out ways to be more politically relevant through their research into minority health, health disparities, and the social factors of health. Through their relations with ethical bodies, scientists enter new territory. They delve deeper into inquiries across disciplines as race finds a higher place on scientists' list of priorities.

Even in cases where studies have not taken race as a starting point, researchers have made sociogenomic links. In one instance, when a Pennsylvania State University cohort found variation between Europeans and non-Europeans in an ancestral gene for pigmentation (which is present in the zebrafish,[115] a model organism for pigmentation because of its striped coloration), they hastened to understand the range of this variation and its meaning for the ethics of race.[116] As two senior scientists of the study wrote:

> One of our most important contemporary responsibilities as scientists and physicians is to use a uniquely human, positive side of human nature—the ability to work together toward idealistic goals, to learn about the history of race perceptions, to accurately represent our new understanding of these issues to the public, and to lead our evolution toward a more egalitarian future.[117]

Consistent with the increasing focus on social matters of race, these scientists issued a call to science activism based on evident biological truths.

A deeper look at how scientists define and conceptualize race will show that a race-positive paradigm and a sociogenomic framework do not necessitate *conceptual* consensus within the field. Rather, as scientists use their own diverse racial experiences to make sense of their present-day responsibilities, they devise subtly different ways to apply a race-positive, sociogenomic perspective to the problem of race. The sociogenomics of race is thus a multifaceted object that carries an abundance of ethical meaning and political valences which will have specific consequences for how society will view race into the future.

4   Making Sense of Race with Values

IT IS HELPFUL to envision the genomics of race as embracing a spectrum of working models.[1] On one end of the spectrum is a "biologically deterministic" model, which corresponds to older typological notions of race.[2] The biologically deterministic model posits race as a proxy for biological distinction. An extreme version is the notion that races are akin to subspecies. Europe's first human taxonomists—natural historians like François Bernier, Carl Linnaeus, and Johann Blumenbach—established a precedent for dividing humanity into mutually exclusive and hierarchically ranked continental subspecies.[3] Subsequent nineteenth-century scientists were so influenced by this idea that they debated whether there was in fact one human species.[4] Eventually, Darwin's notion of races as endogamous reproductive populations—populations of the same species that, through breeding, comprise unique gene pools—replaced the idea that races spontaneously emerged on different continents.[5] Postwar geneticists recuperated the Darwinian notion of races as endogenous breeding populations as a counterpoint to social Darwinist and eugenicist theories that focused on hierarchies of social and behavioral characteristics.[6] Today we can see traces of this argument in some continent-based definitions of human ancestral populations, and in common lay notions of race that assume there are biologically distinct races, which can be typed by a set of visible, heritable traits. It is also present in claims that while race is something biological and social, it is chiefly genetic. When taking such a position, scientists may articulate an array of social factors involved in biological pathways—factors such as diet and health behaviors—but they maintain that the predominantly genetic gene-environment variation falls along the axes of everyday notions of race.

In the middle of the spectrum lies a "weak correspondence" model, which claims that while there are no such things as genetic races, there are socially meaningful groups having significant biological commonalities. At times scientists use this model to suggest that common lay classifications can be suitable proxies for genetic variation. But most often they argue that racial terms come with too much political "baggage" to be of use to genomicists. Many who have mapped genes responsible for skin color, hair texture, and facial features, what some call the "biological aspects of race," have concluded that *admixture* is a better expedient for explaining genetic diversity. An admixture view of human variation retains a belief in original Old World continental groupings akin to the common lay version of race, but argues that each human possesses a mixture of old-world genomic variation.

On the other end of the spectrum is a "social constructionist" model, which may acknowledge agglomerate variation but refuses that a correspondence between such variation, genetic classifications, and lay classifications exists. Social constructionism maintains that race is a product of culturally and historically contingent meaning-making. In many instances, scientists remain agnostic about the true divisions of humankind, yet they do use race to include minorities and tackle racial health disparities, while genomics takes a further look at the reality of race. The far end of constructionism connotes an awareness of sociological processes of racialization—namely, the dialectics of scientific racism and social processes of exclusion, enslavement, segregation, and ghettoization—and a complete rejection of any biological basis for race. This position is often expressed as a radical antiracialism, a refusal to ally biology and social difference, the claim being that there are no real divisions of life such as species, populations, races, or groups. Scientists who take an extreme constructionist position justify their actions by arguing that race-positive science is not only racist; it's simply false.

None of these positions are dominant among the scientists with whom I spoke. Rather, genomicists move between these models as working models, elevating one or another according to the different practical and social concerns that pertain. In fact, because race signifies a set of interlocking notions of difference, its *re*-signification can be witnessed in the space of a single definition, argument, or even sentence. Genomicists make important connections between biological and social rationales, at times in order to challenge common lay notions of race, and at other times to acknowledge and engage public notions of race. It is precisely this flexibility that has permitted genomics, in less than a decade, to nimbly shift from ignoring race to assuming a race-positive frame. While nearly all scientists accept that human variation can be mapped onto the continental categories that earlier racial categories were based on, they avoid reducing race to a single biological or social valence. As a result, race remains multifaceted, contentious, and unresolved, an ongoing question for genomics.

Behind all of these conceptualizations lie some of America's most deeply cherished values about human diversity, equality, and justice. Allowing scientists to think out loud, to finally articulate the concepts that they maintain, further shows how they process meaning around race with subjective—even emotional—registers.[7] As scientists alight on different aspects of racial knowledge, they draw on their own experiences to lend meaning to their pursuit of rationalized knowledge about race, and they apply new understandings to make sense of prior experiential confusion.[8] Such concepts of difference are not just research tools but are also optics that help scientists, as Ann Morning suggests, "make sense of race-coded social worlds."[9] Scientists' multidimensional attempts to define race illuminate how reflexive biosociality pertains to the formation of basic concepts.

Across the spectrum of models—biologically deterministic, weak correspondence, and social constructionist—it is apparent that genomicists with differing views on race have similar values about aiding minorities or creating a more equitable healthcare system. Thus, in addition to presenting this spectrum of views, which takes the form of a polyvalent expression of reflexive biosociality, I examine the *moral order* that can emerge in the face of conceptual instability. Most research on the production of moral orders in science focuses on the ways that common moral ideas shape conceptual *consistencies* and their standardization across institutions and fields.[10] In contrast, I emphasize how a common moral ground can foster multiple, *contradictory* conceptualizations. Scientists with similar experiential reference points, such as life in segregated communities, discrimination met by family members, and intergenerational immigration, simultaneously offer complementary moral rationales while presenting contrasting versions of race.

Genetic Explanations for Race

Frederick, Maryland, is a forty-five-minute drive from the main NIH campus in Bethesda. There, the National Cancer Institute's population genomics lab has replaced most of the military operations that were once responsible for America's biochemical warfare program. The irony was not lost on me as I passed through the northern perimeter's high-security checkpoint. Sipping tea in the cozy sitting room of Stephen O'Brien's office, amidst an array of animal relics, I was reminded that even government scientists who conduct the most basic research have their own stories to tell. The Laboratory of Genomic Diversity exists for the purpose of cancer-related population research, but O'Brien's first love is big cats. After conducting several successful conservation missions on cats in the wild, O'Brien noticed that viruses were responsible for many of the cancers that decimate cat populations. O'Brien's lab applied this oncoimmunological approach to humans and became one of the first labs to find a gene responsible for AIDS resistance and progression.

As O'Brien signed a copy of his book *Tears of the Cheetah: And Other Tales from the Genetic Frontier* for me, he reminded me that humans are only one of many mammalian species that follow the laws of speciation. Comparing feline population dynamics with human migration patterns, he argued that modern continental clusters indeed fit the zoological definition of race as subspecies. He explained that subspecies hold two potentials: substantiality, or the acquisition of favorable genetic adaptations to unique living environments; and discontinuity, the possibility of forming a new species whose members are incapable of breeding with those of the progenitor. O'Brien confirmed that humans possess significant continent-specific immunological responses which could be represented by a racial taxonomy. He suggested that genomics manipulate subspecies variation in order to better fight disease. When I asked him whether the race concept was still relevant to genetic research, he said:

> Well, I think it is certainly relevant to the history of human populations because there has been genetic or geographic isolation, if you will, between at least the major ethnicities: the African group, which has a large amount of the diversity, the Caucasian migrants, the East Asians, and so forth. And when you get down below that level then the question about how much of it is genetic and how much of this is cultural is a mystery. But there certainly are some interesting separations and things that have happened in the different ethnicities which distinguish them from each other and which have genetic bases. For example, we know in genetics that the immune system is molded by defenses in our ancestors that have survived infectious disease outbreaks or chronic diseases such as Alzheimer's or multiple sclerosis or even cancers.

O'Brien suggested that race matters for the anthropological record but also for today's common illnesses. He went on to explain how natural selection produced these continental races, each with its own immunological history. Throughout his explanation, he entwined stories of the "parallels in animals" and the ways that genomic technologies have at last made speciation decipherable. O'Brien's commitment to race as a tool in animal conservation frames his notion of race as a viable and portable biological analytic. In conservation, determining racial taxonomies is the first step toward saving wildlife species. Garnering political recognition of immunological threats is part of the genomic contribution scientists make. Thus, this robust zoological model of race is itself informed by ethical frameworks tied to issues like conservation, environmentalism, and animal rights. It best serves O'Brien to continue to hold a partially typological understanding of race.[11]

Comments like these express the closest thing to a strict deterministic model of race and the rigid, biologically based racial taxonomies of past sciences. They suggest that race has a genetic basis, matches lay notions of race as biological and typological, and fits a continental genetic population schema. Such a rationale could be used to support the production of race-based medicine, forensic profiling tools, and other racial products and services that tailor particular solutions to each racial group. It could also be used to establish group-targeted social policy about race relations and racial justice—just as past genetic discourses of race were used to advance social engineering, immigration policy, and eugenics.

Yet despite voicing such a deterministic view of race, O'Brien also emphasized a vision that interracial mixture would soon overtake the speciation process:

> So, what happens is that you go to places like Rio de Janeiro, and you walk on the beach, and you take skin color as a correlate—there is a continuum which goes from the very, very dark African lineages to the very, very light skin lineages and everybody in between. And, to tell you the truth, these are very beautiful people. They are very attractive and certainly have no aversion to falling in love and making offspring. . . .
>
> Thirty years ago that is the way the beaches in Rio de Janeiro were, and thirty years later there is no majority in California. There is a mixture of many different rich ethnicities that come together. And I think that in the future we will continue to see the mixing bowl, if you will, of ethnicities as long as we continue to recognize that that is probably a good idea.

O'Brien cheerfully called the current continental genetic clusters a "historical footnote in terms of human population [relations]." He thus cast his extreme genetic definition of race in terms of a brighter postracial future just down the road. To O'Brien, mixing is a "good idea," and a sign of the power of social relations to shape evolution. In further conversation, O'Brien used the term "major ethnicities" synonymously with race to connote this unified biological-cultural process. His views convey the awareness many genomicists have of the negative uses of arguments about distinct biological races. Contrary to an objectively rationalized definition of race, they communicate the normative belief that the world would be a more harmonious place without racial divisions. Genomic elites see themselves as shepherding society toward that future with their clearer understanding of evolutionary dynamics and the good news that race will soon fade away.

Just down the hall at the Cancer Institute, Mike Smith expressed the more pessimistic view that race will likely be of continued relevance to medical research for some time to come:

> SMITH: I think that race is a very touchy subject. I wish it were not a scientifically useful construct because it'd make it a lot easier if it were not. We have to take race into account all the time when we do our genetic analysis, because if we don't, we miss things.
>
> BLISS: And by race you mean . . . ?
>
> SMITH: When I just used *race* in that context, what I meant was splitting people into groups of, let's say, European Americans, African Americans. Hispanics get to be a little messier—race is not really appropriate there.

Smith attributes race to patterns in genetic variation. He equates race, continental groups, and genetic populations, all based on the idea that our common parlance about race provides an obvious shared meaning. Like O'Brien, Smith communicates the ambivalence scientists feel toward the concept, acknowledging that it carries a lot of political baggage which the genomicist must manage.

Though he would rather not have the responsibility of deciphering such a politically charged term for the public, Smith sees genomics as best poised to deliver knowledge about race:

> It requires a lot of knowledge of the populations under study, and then appropriate markers to, at least in the current state of the technology, to make that call. Is this an African chromosome? Is this a European chromosome? Is this an

Amerindian chromosome? And so, obviously, if you are an African American, you do not need to worry about the Amerindian. We know that that [mixing] is not a major issue, so you don't have to look.

By this definition, race is a composite of continental ancestry void of specific political meaning and only knowable by the most cutting-edge technologies. Statistical properties and historical patterns in mating tell scientists where to look for chromosomal diversity.[12] Traditional racial categories form the backdrop of genomic medicine.

Still, in explaining race to me, Smith did not avoid reflecting on history, politics, and science's social impact. He is concerned that scientists have the choice to illuminate or ignore the facts, and worries that scientists will choose not to study race out of a colorblind logic. To convey his disdain for ignorance of racial inequality, he explained the impression that segregation made on him as a white youth in Aiken, South Carolina. Smith said that though Aiken had a steady stream of migrant Northern laborers working in its plutonium plant, the town itself was racially bisected by a set of train tracks: "The kids on the other side of the track were definitely *black*." Smith's struggle to make sense of segregation led him to be critical of those who chose to look the other way and who offered superficial, semantic solutions to racial problems. Smith's thought process shows that as with lay perception,[13] scientific perception of race draws on racial experiences and memories which are themselves pregnant with meaning. Scientists make sense of their past experiences with their science, and make sense of their science with past experiences.

Smith viewed my suggestion of replacing racial classifications with the term "continent of origin" as ineffectively thin, for its failure to capture the full historical associations that Americans live with in dealing with race. As he explained:

> Maybe it's just me, but I think that if what you are dealing with is basically race, why not call it that and then live with the consequences of people that might say, "That's inherently racist." . . . But to call it a continent of origin and to mean race?
>
> That reminds me of a story when I was a kid. My parents show up from California in Louisiana, and there are three bathrooms: one, two, and three. And I am four [years old] and I go to my parents and say, "I don't know which one to go in; they're numbered." And three was for the blacks. You got a different label, but it is the same end result.

Smith's pessimism derives in part from the shell game of racist symbols and monikers he has witnessed in his lifetime. He thinks it ludicrous to believe that changing scientific terminology alone will eliminate racism or racial beliefs. In fact, like most of his colleagues, he believes that helping minorities requires using race as a guide to biological information: "We have projects in the laboratory that are examining HIV progression in African Americans. And we are taking the amount of admixture in those people into account. Because it may be an important quantity and it may also be a way to help us find genes behind the HIV progression—even more importantly, behind HIV infection." Remarks like these signal the way many scientists approach race as an instrumental tool in a larger social justice struggle for minority health. For them, utilizing race in the lab is both scientifically sound and ethical. To scientists like Smith, race-free genomics is the same as the colorblind rhetoric that contributed to racism in the South. Like all those I spoke with, he conceptualizes race as biologically and socially meaningful in ways that demand immediate attention from genomics. Sentiments like these betray the way scientists envision the field's role as both a corrective to the prior socially constructed scientific engagement with race, and an arm of a broader racial consciousness-raising front.

Three thousand miles away in San Francisco, I spoke with several University of California scientists who were involved in the launch of the university's Center for Pharmacogenomics, and who routinely use race synonymously with genetic population. Their comments best exemplify the common belief that race is a mélange of biological and social factors, in which genetics is crucial. It follows that genomics must incorporate race into its protocols. At Esteban Burchard's lab, only a stone's throw from San Francisco's Mission District, the predominantly working-class Latino community where he grew up, Burchard spoke of the difficulties of coming up with operational categories that represent all the diversity scientists want to study. He told me: "Race is a complex construct. It includes social factors; it includes self-identity factors; it includes third-party factors of how you view me. But it also includes biological factors." Burchard explained that gene-gene and gene-environment interactions affect minorities—a combination of genetic heredity and social standing. He wants to maintain a flexible working model that accommodates the genetic, biological, and social aspects of race. In line with the sociogenomic approach, he believes it is his duty as a scientist to study all the factors affecting health.

Burchard calls his research into race his "personal passion." Along with one quarter of the scientists I interviewed, Burchard said that his interest in minority health and genetic admixture stems from his own experiences living in a multiracial family.[14] Burchard is particularly concerned with the way that racism has created different life expectancies and opportunities for his Latino forebears, his Amerasian children, and his Southeast Asian wife. Looking at a photo hanging above him, he said:

> My mom was really hardworking, and my grandmother also worked very hard. She was very smart, and I always thought—I still think to this day—that she didn't get her due credit because of her racial background. I still believe that if she was white, she would have been much more successful, but she looks black. If you saw her, you wouldn't think that we were related.

He told how his mother predicted that his lighter skin color would gain him entry into worlds closed to her:

> And she told me once when I was very young—I must have been like seven— she said, "You are going to be able to go places where I could never go." She taught me about this permeability. I use it today. "Because of who you are and your abilities, you can go through certain scenarios back and forth, kind of like a fish going through a big net. Some fish can go back and forth while other ones will get caught."
>
> And she told me very young that I would be able to do that. And she was right.

Burchard feels a responsibility to foster a "by the people, for the people" science, which his mother and grandmother did not have access to.[15] Preoccupations with biological kinship condition his efforts to create a Latino genomics program.

Burchard also gestured toward contemporary racial experiences and how they sway genomicists' thinking about race. Working in an environment with little racial diversity, he told me, is a daily reminder of the racism that affects all members of society. This motivates him to prioritize minority health and minority recruitment in his lab, even in the faculty and staff he elects to work with. His applications of admixture mapping technologies to Latino populations have made him one of the most influential and visible scientists in the field. Burchard cultivates what Kwame Anthony Appiah has called "living *as*": he goes beyond identifying with a social identity-based group to behaving in ways that solidify society's basic concept of that group. All the while, as a racial

insider, he is creating a sense of credibility about the biological and social ins
and outs of race.[16]

Burchard's collaborator Elad Ziv expressed a similar conception of race
built on a single classification system for social, biological, and genetic factors.
Ziv stresses the need for genomicists to match their taxonomies with those in
the larger system of public health,[17] while maintaining that this makes sense
genomically. At Mount Zion Medical Center, he told me:

> This is the data that [epidemiologists] have. This is also how people think of
> themselves to some degree. So you can get this information, and it turns out
> that it correlates with a lot of diseases, for various reasons, as a functional thing.
> You see a lot of associations with diseases—pharmacological benefit or risks or
> side effects—which is kind of what is really interesting to geneticists these days:
> to see if we can sort of find those causes of type II diabetes, cancer, asthma, and
> what have you.

Ziv sees pharmacogenomics as the key to solving the most pressing health
issues, and race as an integral tool. Alluding to the shifting terrain of racial
identity, he acknowledges a deviation between genetics and "how people think
of themselves." Yet he believes that there is enough correspondence between
the "real," underlying biological factors and the socially constructed top layer
of identity to merit using race as a proxy. While "cultural factors" of race
are woven into such epidemiological rationales, this model insists that life-
threatening biological and genetic factors are the most urgent. Speaking about
cancer, Ziv remarked:

> Under thirty-five, it's rare in all groups. That's good. But nevertheless, when
> we see it *double* under age thirty-five, I don't think that that's a sociological
> phenomenon. I think there's something going on there. I can't say for sure, but
> I think it's biological, I just *know* it's genetic. It could be exposure to something
> that's bad, but if it's possibly genetic, then I think it's worthwhile studying.

In this sense, the "something going on there" is a gap in racial understanding
that genomicists are best poised to fill.

In a display of field-building empathy commonly taking place as scientists
think about the problem of race, Ziv told me that he believes science activism
is responsible for important changes in public understanding of race. These
efforts include presenting genomic findings to new audiences and linking

genomic inquiry to political needs. Animatedly describing the effect Burchard has on people, he exclaimed:

> [Esteban] gets people really fired up when he goes to talks! He says, basically, "This is NIH! And this is *your* tax dollars at work studying *your* diseases. And if you think that basically your group doesn't need to be studied, then you're just wasting your tax dollars. So, you should be out there! You should be promoting and encouraging whatever group you're from to be studied, because this is *your* money."

Ziv agrees with this "very sort of utilitarian" or "bottom line" way of looking at minority health. Echoing his article in the *New England Journal of Medicine*,[18] he argued that genomicists who wanted to be politically correct had promulgated a "race-neutral approach." Like Smith, who critiqued colorblind science, Ziv views race-neutrality as harmful to minorities because it leaves their health concerns out of the picture. Though acknowledging that he is slightly uncomfortable being in the spotlight for his race-positive views, he champions the genomicist's proactive public responsibility:

> I think it's actually a good thing when people are actually thinking, "Well, why is it that this group gets this disease?" Why is it that, for example, Southeast Asians get type II diabetes when their [body mass index] is actually 24 or 23? They're skinny, but they're getting a *horrible* type II diabetes. I think that it's a good thing that people think about why African American men are getting more prostate cancer than any other group.

For Ziv and many others, the first step is for scientists to raise consciousness about disparities in disease rates. At the same time, Ziv draws a clear line between scientific and public understanding; he claims that genomicists are more comfortable than nonexperts with a multifaceted biological idea of race. His challenge to colleagues to get more involved in changing this public misunderstanding shows that biologically deterministic models for race can go hand in hand with a liberal activist ethic.

Up at the University of California, San Francisco's Parnassus campus, Pui Yan Kwok explained race in terms of continental genetic clusters and gene-environment biological factors, but he also talked about cultural know-how. When I asked Kwok to clarify the genetic population–based explanation of race, he highlighted his everyday subjective experience of race: "For some races, you know, I can't tell them apart except for the continental stuff, but for other races I can tell their [ethnicity]. Like in Asia, I can tell if someone is from Japan versus someone is from Korea." Kwok explained that when he first came to the United States, he didn't register the same racial cues that Americans did. Hair color and eye color were less important to him than "whether the face is rounded or tapered." He literally trained himself to see pigmentary differences and associate those with the racial categories he uses in his science. Kwok assumes that there is a biological basis for race but that phenotypic distinctions are not always to be trusted.

The cultural variability of racial meaning was commonly articulated by scientists born outside the United States. Foreign-born scientists like Kwok spoke about educating the public about biological and social diversity. Kwok went on to describe his own education in race: "When I was growing up, even in different regions in China, people were, like, 'Oh, I'm from Shanghai and so I'm more sophisticated.' I think the race thing is kind of mixed into that. People think that certain ethnic groups are stupid or lazy or things like that." For Kwok, stereotypes and cultural assumptions are built into the ways people interact, thus imprinting health behaviors and the life course in significantly racialized respects. If genomicists can alleviate the stereotyping, they will have done a good job.

Like O'Brien, Kwok reflected the view that the world would be better off without rigid notions of race. He referred to his own loosening of boundaries:

> You know, racial profiling; we do it every single time. And I catch myself doing it. But then sometimes it's almost like a *protection* behavior. It's like when I go visit Asia, there are certain people that I just try to avoid . . . but the good thing is that I also get to know someone, and the stereotypes break down, and then you can interact in a kind of good way.

Kwok described race as a "shorthand" that leads to unnecessary prejudice but also as an unavoidable way of perceiving the world—"When I go out to the streets of San Francisco, I can tell people's races." At the same time, like all others I spoke with, he emphasized that genomics will one day prove the different racial preoccupations produced by cultural contexts to be insignificant. The concept of race Kwok offers rests on a belief that racialism is both genetically founded and culturally salient—biologically hardwired yet socially fluid. There is no universal, commonsense taxonomy. Kwok hopes that, possessed with new genomic knowledge, people will think twice about allowing their stereotypes to guide their behavior.

The definitions of race given by these scientists suggest that genomicists who offer genetic explanations also draw on constructionist themes and social considerations when relating their fundamental understanding of race. As we analyze scientists' articulation of their racial beliefs and as we map currents of essentialism that underpin those beliefs, we should acknowledge the depth and range of their rationales. Just as varying types of essentialism may underpin a single explanation of race, and even feed into one another,[19] different rationales can engender a common moral ground. The scientists here often justify their claims by casting them as race-positive, as opposed to colorblind. They describe using race in order to include racial minorities and target minority health. These themes permeate their various articulations of race, thus fabricating moral cohesion from diverse understandings. Scientists who perceive a genetic basis for race are no less concerned about the social ramifications of race than their colleagues who prefer a more constructionist understanding.

Conceptualizing a Weak Correspondence Between Genetics and Race

In contrast to the metonymic assumptions of the positions exemplified in the previous section (that is, race as proxy), genomicists also work with a model that draws a nominal line between racial categories and what are currently held to be humanity's major ancestral populations. I say "nominal" because scientists still maintain that these groupings look very much the same, because they are both defined continentally. A model allowing only a weak correspondence between genetics and race permits scientists to advance an antiracist racialism that critiques the language of race. Meanwhile, genomicists can use the same research variables that scientists have been using throughout the history of human biology—those of continental populations.

At Palmer Commons, on the central campus of the University of Michigan, I met with biostatistician Noah Rosenberg. A coauthor of the famous 2002 *Science* study on Diversity Project samples (Chapter 3), Rosenberg has appeared in the pages of *Discover* and *Wired* for his expertise on hominid speciation. He is also famous for some of the field's seminal quantitative methods for determining the number and nature of ancestral lineages; for example, the "informativeness" quotient for gene variants, which is used in population determination. Rosenberg explained the view that race and ancestry correspond closely:

> There are overlaps between the different concepts in the sense that if you subdivide people by genetic analysis, you get subdivisions that correlate with

subdivisions that you would obtain from, say, the U.S. Census racial categories. They correlate but are not identical. . . .

So when we have talked about clusters, the cluster is an abstract statistical object. It does not correspond exactly to a geographic region, but it correlates quite closely with geographic regions. So not every Eurasian is 100 percent in the cluster that corresponds to Eurasia. Some of the individuals have a partial membership in the cluster for East Asia.

Rosenberg's coauthor Jonathan Pritchard agreed. One of only fifty-six scientists selected for a multimillion-dollar Howard Hughes Medical Institute Investigator Award, Pritchard has fast risen to become a leading expert on evolutionary selection and gene regulation in genetic populations. His STRUCTURE software has been at the base of a great number of population genomic articles discussing the biological value of race, and his research on natural selection has been featured on the front page of the *New York Times*. Pritchard is also a scientific advisory board member at 23andMe. At his University of Chicago lab, he stated:

You can subdivide human populations according to genetic information at lots of different levels. But if you do clustering with about five, six, or seven different groups, then those groups correspond pretty well with continental ancestry, and the groups you get correspond pretty well to popular notions of race. So we don't usually use the term "race" in our scientific writing, because it is kind of imprecise and different people mean many different things by it, and also because the term comes with a lot of baggage. It's just not really very helpful. But the clustering that we get out of doing cluster analysis here does sort of correspond pretty well to what many people think of when they think of race.

In these scientists' terms, only genomic computation reveals the true patterns of ancestry, yet race and genetics are undeniably correspondent. In other words, genomic software confirms the continental ancestry that natural historians like Carl Linnaeus and Johann Blumenbach imagined could be grouped on the basis of morphological commonalities. Pritchard and Rosenberg acknowledge that racial terminology is politically problematic because of its exploitative application in history. Thus, they argue for a nominal shift: substitute "black" with "African," and "white" with "European." This division upholds the pillars of common typological notions of race while attempting to avert their political significance.

Like the scientists in the previous section, Pritchard and Rosenberg draw our attention to the genomicist's role in shaping public thought about race. They don't deny that genomics should be sensitive to matters of terminology, and are aware of potential pitfalls in public interpretations in the face of a nominal shift. Pritchard believes that genomicists are responsible for their incendiary findings: "So I think that it is important that if you got something that is inflammatory, but there is a 50 percent probability that it is wrong, then it's incumbent on *you*—you have to be really careful of how you present that." Pritchard said that he and Rosenberg extensively deliberated the best way to present their study's "single clear example of genetic clusters corresponding to [race]." They reviewed how other racially significant findings were publicly received, and even questioned whether they should publish the findings at all: "I mean the way that we wrote the article was kind of motivated a lot by this sort of concern of it being misused by people who want to think that genetic differences between races are more important than they are." In the end they decided to release their findings, but "not exaggerate the importance of them." Pritchard felt that they had done a good job and that the fallout was minimal—though other scientists I met were less positive about the response they had gotten to similar declarations of categorical correspondence.

Scientists often temper the weak-correspondence argument by emphasizing that ancestry should be conceptualized differently when thinking in anthropological versus medical terms. At Stanford University's Gilbert Hall, I spoke with Marc Feldman, who argued that a continental register is more appropriate for anthropological analysis, whereas medical categories likely require a higher-resolution variable. Feldman claimed that though genomic cluster studies have produced "ancestry clusters essentially by continent . . . those clusters defined by us have not been shown to have *anything* to do with medicine. To do that, you would have to have a much deeper knowledge of individual environments, and it may be that you really need to do that genetic epidemiology on an incredibly fine scale."

Like others, Feldman has a deep desire to trouble lay notions of race, and a proven history of doing just that. Feldman gave the example of sickle cell anemia, a disease originating in the equatorial regions of many continents: "Identifying somebody with hemoglobin S doesn't mean that they have African ancestry," he argued. Invoking the distinction between phenotypic and genetic populations put forth by postwar evolutionary synthesis scientists, Feldman concluded that the phenotypic classifications that have reflected prior notions of race have not been good indicators of medical similarity and that "typological features and the genes that determine them are not characteristic of the genome as a whole."

Thinking about the bigger picture of genomic knowledge about race, Feldman drifted to entanglements of the scientific and the social. After explaining the most salient social distinctions of his youth in Australia, and their clear cultural origins, he said:

> [I]n the university situation, you today are forced into acknowledging what [race] means because of affirmative action . . . and the fact that we have so few minorities on the faculty, and so few coming up to be on the faculty in the next generation.
>
> If you do demography like I do, then you [also] have to continually focus on what's the difference in life expectancy and morbidity between the different self-classified racial groups within the United States. There's an article in today's *New York Times* about that life expectancy and how different it is among the different minorities—Hispanic, Asian, African American, and Caucasian or European-based.

From the affirmation of continental genetic clusters to racial hiring practices to health disparities, Feldman ran the gamut of considerations invoked by racial sense-making. As with the other scientists, questions of race provoke him to carefully weigh the scientific and social responsibilities he holds dear. Feldman's words underscore the common belief in the field that there needs to be more minority recruitment and, further, that there is a connection between having minorities on staff and having a rigorous minority health agenda in genomics. Along with many in his generation, including those advancing a biologically deterministic model, he thinks through the segregation he sees in his lab by reflecting on segregation in society and health inequalities.

Aware that defining ancestry in continental terms does nothing to challenge the basis for race-based therapies and a race-based social healthcare system, and that maintaining that society is made up of biologically distinct groups can justify arguments for a "separate but equal" social order, scientists also temper the weak-correspondence argument by fashioning alternative signifiers. Luca Cavalli-Sforza, for example, has devised an alternative way to envision genetic variation that he believes can sidestep rigid notions of continental races:

> When I refer to the major clusters, which are six or seven, basically I refer to them as *pseudo-continents*. First, I would say they have similarity, but really they are not continents. Basically, the major difference with respect to the standard is that Europe, West Asia, and North Africa are more similar to each other. So that should be a pseudo-continent.

The boundaries of Cavalli-Sforza's pseudo-continents slice right through our commonly held notion of genetic groups, or clusters. He bars the use of lay racial categories in his Stanford lab, because he feels they cannot adequately convey the complexity of the human record. Yet his notion of pseudo-continents continues to reinforce common lay understandings of race in invoking ideas of racial divergence and subspeciation.

Though time and again I heard scientists like Feldman and Cavalli-Sforza say that medically relevant clusters do not necessarily match those in evolution studies, when discussing medical genomics many routinely invoked continental clusters as the basis for disease research. At the Broad, where David Altshuler painstakingly argued that race was a socially constructed classification system, he yet expressed the collective view that continental comparisons are critical to genomic statistical models:

> If your ancestors lived in Africa from, like, 10,000 B.C. until 1800, something like that, or 1500, then there is a measurably distinct allele frequency pattern than [that of] people's ancestors in that same period who lived in Europe, within Asia, within the Americas and Pacific Islands. I am not saying that there aren't any variabilities within each of those groups, but the variabilities within are much less than the variabilities across, and of course, [as] has been known for many years, most of variations everywhere.

Altshuler's formulation of the "more variation across than within" statement is a reversal of Richard Lewontin's "more variation within than between" phrase, which was turn-of-the-century academic common sense.[20] Though genomicists at the close of the Human Genome Project largely agreed that the elucidated genome proved that there was more variation within races than between them, genomicists have recently qualified that statement with indications of differential population frequencies for continental groups. While diverging from a biologically deterministic concept of race, this model still supports a continental framework that could potentially be used to argue for race-specific drugs and biotech applications, not to mention race-based health programs and medical welfare.

At the same time, scientists like Altshuler describe race as a confounder of underlying factors that, taken together, merit further analysis, "because a label can itself be the cause of that health outcome, it could be a source of discrimination or of behavior or culture, or it could be genetic." Like the scientists who define race genetically, Altshuler calls for genomicists to walk this fine line, taking on more health disparities research while developing better methods for understanding the major patterns of genetic variation. Though he claims that genomics—even anthropological strands—should steer away from taxonomies that even remotely smack of common lay notions of race, his argument portrays the field's commitment to manipulating continental ancestry in order to bear down on racial health inequality.

In the articulation of the weak-correspondence model, dedicated as it is to replacing racial terminology with geographical terminology, we see a thought process motivated by many of the same concerns evident in biologically deterministic explanations. These scientists also connote the desire to ferry society to a world without racial distinctions and to create a higher moral ground for practicing science. Though some scientists proffer definitions that require a flexible use of categories between the medical and anthropological sides of the field, these concepts—just as with deterministic rationales—rely on experiential knowledge and a concern for political equality, which pushes these scientists to consider a strategic use of race.

Genomicists from all corners of the field are apt to offer what they see as the biological aspects of race, even as many attempt to refute the existence of genetic races. Many do the former defensively, because they are worried that ignoring the congruity between ancestry and race will produce disbelief and cynicism in the public. This was an especially common theme in my conversations with scientists who self-identify as Jewish and have day-to-day thoughts about which biological differences should matter to society.[21]

As I sat with Joel Hirschhorn in his office at the Broad, he recounted his family's experiences with the Holocaust, the biological significance of Jewishness in the context of Nazi "racial hygiene" and American eugenics, and the current import of these concerns to population genomics. As he moved from the historical details to how he deals today with the weight of the intellectual legacy, Hirschhorn showed me a recording of a lecture that he gives on scientific racism at Harvard, and he exclaimed:

> There were these people who were measuring skull cavity sizes! And trying to show that their preferred order of races correlated with increased—the better [the] race was, the larger the skull size. . . . But there was actually a guy, Agassiz, at Harvard, who sort of took to this and really publicized it. And he was actually an incredibly well-known scientist, sort of like a rock star of science! He was a *geologist*. He discovered there was an ice age!

Hirschhorn was shocked and dismayed to learn about the ease with which scientists in his own institution had traded on their reputations to make racist claims. He has taken it upon himself to caution Harvard's future elites about the responsibility that attends the practice of human variation science.

Yet Hirschhorn is cautious not to overstate the irrelevance of biology to race. Today's scientists retain a fear of this legacy, he remarked:

> Now, there are people who go incredibly far in that direction, which is to say not only is there no such term as race, but there is no biological reality that even correlates with race. That's where we differ from those groups. . . . [If] you took a bunch of people—and you lived in the United States; it would come out differently perhaps if you did this in other countries—and you said, "Okay, I want you to assign races to all these people," and then you took the genetics, the genetics wouldn't be perfect, or the racial labels wouldn't be perfect predictors for the genetic ancestry, but they would certainly be correlated.

As with the other scientists, Hirschhorn assumes a correspondence between race and ancestry. He averred that the genomicist's responsibility is to ensure that their science will not be used to "justify mistreatment," but he added that being dishonest about biological correspondences is tantamount to colorblind science. Hirschhorn reiterated concern over the public's preoccupation with visual, morphological cues; genomicists are in a position to shed light on variation in a socially conscious way. His comments bespeak the subjective and political roles scientists see for themselves. Because they believe that genomicists are responsible for the social implications of their research, scientists like Hirschhorn carefully attempt to sort the "biological reality" from the social conceptions and misconceptions.

David Goldstein, coauthor with Hirschhorn of a *Nature* article criticizing lay concepts of race, also commented on the different genetic frequencies in racial groups: "My current working model is that racial and ethnic groups are *not* the same [from group to group] and gene variants *can* have different frequencies." Goldstein professed that there are genetic groups to be found, and that ethnoracial groups do possess differences which are decipherable by genomic technologies. Still Goldstein communicates the understanding that such differences may not have functionally different effects: "But if they're there, [genetic differences] probably will have a similar *effect*, because we're physiologically similar." Goldstein reasons that if the outcome of genetic variation is the same across racial and ethnic groups, the variation cannot constitute a medical priority.

Over drinks at a café near his lab, Goldstein continued:

I consider race and ethnicity as relatively poor guides to the structure of genetic variation. But that is a really different thing from saying that different racial and ethnic groups are all [the] same and that the racial and ethnic labels are uninformative about the pattern of genetic variation. To me, that's always been abundantly clear. . . .

I don't like using race and ethnicity, because it's not good enough for what I'm interested in. What I'm interested in is representing the pattern of genetic variation on the planet, and I can use geographic ancestry for that purpose well.

The aporia expressed here, what Goldstein terms the "glass half full/half empty" dilemma, was voiced by a range of scientists articulating a partly constructionist position. On the one hand, they want to remain sensitive to people's need to match their social identity with genetic evidence—something many now experience firsthand. They believe genomic technologies can point to real ancestral variation amenable to that need. On the other hand, they desire a reclassification of humans that would be based on the genetics of drug response or other medically valuable information. "The best guide is I take a mouth swab from somebody and I genotype them," said Goldstein. Social classifications are thus to be superseded by genomic data. Scientists work amid this conundrum, and as Hirschhorn put it, they attempt to "walk gingerly" over the "landmines" of genetic ancestry, personal identity, biological features, and social affinity to give the public a better picture of the real diversity within.

Speaking with admixture specialists across North America, I heard genomicists explain race as an imperfect proxy for genetic ancestry, for people's racial identities mask hidden continental lineages. Exposing admixture goes hand in hand with the field's revolutionary hopes for its technology. At Harvard Medical School, David Reich argued for entirely replacing the concept of race with that of ancestry. Reich's principal components analysis program, EIGENSTRAT, is one of the field's go-to ancestry estimation technologies, and his lab has consistently made science headlines for its seminal applications of genome-wide admixture mapping technologies and studies of hominoid speciation. Reich explained ancestry as something between racial self-identity and genetics: "I know 'race' is a word that has a lot of meanings, and so I guess I and many of my colleagues think about ancestry because it is not as loaded with meaning. When we think about ancestry, it is often self-described ancestry with lots of people in this country." Like Rosenberg and Pritchard, Reich presents a nominal shift toward classifications that avoid political dispute. His comments remind us how scientists defer to the public's ethical concerns, as opposed to ignoring them or dismissing them as unscientific.

Admixture scientists maintain that continental clusters are indicative of the major human migrations and patterns of gene flow, so they allow research subjects to select their ancestry from a list of continental groupings. Still Reich explained that continental mapping admixture actually pertains to specific *pieces* of DNA: "The way I think of that is not race—but if you trace the ancestry of that bit of DNA back forty generations, did that bit of DNA live in someone who lived in Africa or Europe, or was it in a Native American living in the Americas?" Comments like these deem that neither race nor continental ancestry fully describe the complexity of genomic patterns in a given individual. Yet they also reify the belief that continental variation is a core axis of genomic diversity.

Remarking on the gravity of dealing with others' self-identities, Reich pointed to the potential for genomics to conflict with self-identified race:

> [T]he findings about ancestry can sometimes be at odds with what people think about themselves. It's not particularly the case for African Americans, who I think have a pretty accurate sense of their history, but it may be more the sense of African Americans than other populations. The findings, more generally, about human populations outside of the United States can be described with people's sense of who they are and what their history is. So, it's a responsibility you have to take very seriously.

Reich wavered between the ideas that African Americans knew more, or less, about their ancestry. Like all the others I spoke with, Reich is especially worried about how to communicate best with African Americans, given America's sordid history of racially abusive medical research. Reich said that he tries not to let his "amateur sense of what the issues are" stop him from pursuing medical leads, but that he aims to be especially sensitive when publicly revealing what he and his colleagues see as the real genetic underpinnings of self-identified ancestry in minorities. Taking the politics of identity into account is thus a requisite of genomic research and another opportunity for scientists to consider their own values about race.

Mark Shriver also emphasized how genomic admixture studies trouble lay notions. Since Shriver is now notorious for his controversial forensic and commercial ancestry software, which predicts what criminal suspects look like on the basis of their DNA, he takes every chance he can get to disassociate himself from typological notions of race. After a Friday lab meeting in which graduate students presented on the merits of race-based medicine, he told me:

> People call the Yoruba an African population. It's possible of course for people to say, "well, that's an African" and "he's an African." "They're all Africans and they're the same." This homogenizing across a population is what people *imagine*, and it's definitely compounded when it's a metapopulation that you're imagining is homogeneous.
>
> You come at it again when you get to the New World, or in other parts of the world where there has been admixture, by *that* being ignored as well.
>
> So African Americans are not only variable because they come from many parts of West Africa, but they are variable because they have different admixture levels!

Shriver attempts to problematize the notion that continental populations exist. In Africa as well as the Americas, according to him, admixture is the norm. Like Reich and others, Shriver uses African Americans as the exemplary genetic population—one marked by exogamy as much as endogamy, and one with political sensitivities that should be the baseline of genomic ethical frameworks.

Shriver attempted to explain to me the relationship between the genetic and the social components of race, emphasizing the socially constructed nature of scientific measures. Speaking of the continental reference populations from which ancestry informative markers are derived, he remarked:

> If you searched Africa carefully enough, you could find markers like [the Duffy antigen allele], which really are almost homogeneously distributed at the same frequency across the whole continent—almost. If you focus on just those markers, which are pretty much what we have done in putting this kind of test together, then you can go back to continental terms. But, that's not saying those continental areas are *real*.

Shriver expresses a common belief that the public is "smart enough" to understand the heuristic value of continental comparison and to accept that the tools genomicists use are sometimes "not real." Still he and others acknowledge that admixture mapping technologies are used to develop interventions for continental populations. For example, an admixture study that teases apart African and European ancestry is conducted to find candidate genes in one or another portion of a chromosome so that targeted therapies can be produced. Therefore, though this concept goes against any rigid biological or constructionist concept of race, it contains the seeds of a continental system that can be put to work in a racial framework. For this reason, Shriver's forensic innovations have been received as racial technologies and deemed racist by many social scientists, bioethicists, and even fellow genomicists.

As Shriver contemplated the implications of common genomic models of race, he became uneasy thinking about past failures in science communication and the criticism he himself has garnered. He noted the trouble bygone biologists must have had in friction with public views:

> There are probably some things that they were doing right—I don't really shy away from it. There are politics, too, involved in this. And I think a lot of the problems, though, were in the interpretations and the political uses to which that kind of work was put. That fed back into what kinds of projects were being funded.

Unprompted reflections like these typify the confusion many scientists feel at the way their work has been publicly perceived and applied, despite their efforts to dissociate their own classifications from common lay taxonomies. Moving on to his rationale for continuing to fight for a better understanding of human variation, Shriver referred to his interest in human diversity as spiritual. Like many others, he interprets his work as not just ethically but morally helping to draw together the medical and anthropological aspects of human evolution to set the field on the correct course with respect to race and human variation. With minority inclusion and targeted medicine as his priorities, Shriver commits to create a common moral foundation that can sublimate prior scientific wrongdoings.

As we spoke at DNAPrint headquarters in Sarasota, Florida, Shriver's former business partner at DNAPrint Genomics, Tony Frudakis, also wrangled with what he sees as public misunderstandings, but he still confirmed the correspondence of categories in some cases:

> The idea of race is *highly* imperfect. It's subject to arbitrary [interpretations] in many cases, so clearly it's not going to correlate perfectly with much deeper and biologically relevant measures of population structure. Largely, you find, though, that people that type with African ancestry call themselves as members of a certain group, or a certain set of groups.

Here Frudakis articulates a common problematic in the field. On the one hand, genomicists see recent admixture between Europeans and Africans as a history swept under the rug and best exhumed by genomic technologies. On the other hand, they believe America's "one drop" rule—the norm that posits people possessing "one drop of black blood" as black—creates a greater correspondence between black self-identity and genetic ancestry. This has important consequences for race-based medicine, as it implies that "black" might be a better proxy for ancestry than other lay classifications.

Frudakis illustrated these tensions by showing me the literal genomic breakdown, in print, of his own family members. Indexing the Y-chromosomal and mitochondrial DNA color-coded pie charts of each person, he said:

> Well, I'm Greek. My dad's Greek. Actually, my mom is just run-of-the-mill European. My wife is Mexican, and we have three children. This is me, mostly the blue color European, a very small amount of Native American. This is my sister, my dad, my mom, and this is my wife. So, you will note, first of all, that our three children are about halfway between, but you will note that each one has their own [profile].
>
> It's interesting that [my wife's] dad was a Hispanic, but he had dark, kinky, curly hair and a very dark complexion for Mexicans, much darker than the average for Mexico. No one ever really talked about "what is your race?" He was just always Mexican, he is Hispanic, and he had considerable African [ancestry]. . . . You'll notice that each of her sisters have significant African ancestry, and the mother has none.

Frudakis is yet another scientist who personalizes his science, not simply musing about the potential for his applications to help people learn about the complexities of race but rather testing his own innovation against his own family's racial history. Frudakis insisted that ADMIXMAP, the admixture mapping software that his company was based on, would eventually contribute a truer picture of the diversity within cultures and families. As with a mantra held dear, he concluded: "When people are together, they breed; when they are mixed together, they mix." This belief, expressed by scientists ranging from the most deterministic to the most anticlassificatory, connotes the hope for the eventual fading of continental difference and crudely targeted race-based applications.

The weak-correspondence model suggests that even when genomicists offer a more nuanced view of race, the common concerns of public accountability and desire for racial equality continue to influence scientific thought about race. Scientists would like to move beyond the category of race, but they do not want to obliterate opportunities for research into health disparities or discussions of racism. Their ethical concerns link them across heuristic divides to others who would also criticize colorblindness. Such concerns form but one component of the moral order that genomics presents.

Committed Constructionism

Despite the genetic valence of most discussions, not a conversation passes without the genomicist highlighting the social aspects of race. Listing factors like culture, health behaviors, diet, and environment, genomicists avow a responsibility to study racial health disparities whether or not genetic causes are primary. In the garden outside the headquarters of 23andMe, Joanna Mountain described race as the "undefined variable." Mountain worked at Stanford on African population genomics before establishing 23andMe's research platform. Though Mountain sees a distinction between race and genetic ancestry, and thus believes race can only ever serve as a proxy for other biological variants of interest, she expresses the shared belief that it makes no sense to reject it out of hand:

> There are genetic differences in terms of frequencies between these socially defined categories of race, and therefore when people see correlations between racial categorization and health disparities, you cannot investigate them without continuing to use the term *race*. It is irresponsible to ignore them, I think, because I feel they are *dramatic*. Some cancer rates are quite different. And progression rates, whether they be social [or genetic]—there are biological differences.

Mountain most clearly articulates the genomicist's antipathy toward colorblind science. While her social constructionist working model is different from say Feldman's "anthropologically interesting yet medically unnecessary" or Goldstein's "functionally equal despite varying gene frequencies" characterizations, it is motivated by the same sense of responsibility that fuels scientists to tackle health disparities and create a socially equitable science.

Mountain echoed the University of California scientists earlier who maintained that nongenetic biological factors are just as important to health as genetic ones: "But biological difference is not necessarily genetic! And that is something that I often find the people are not very clear on. They think biological is the same as genetic, but biology comes from all kinds of things." Like most, Mountain chalked up the unresolved issue of what scientists should do with race to lack of expertise and to misunderstanding. She hopes her work will help the public understand the heterogeneity of race.

Like Kwok, Mountain also discussed at length what she sees as a ubiquitous social perception of racial difference, by talking about her own racialism. Mountain's service tour for the Peace Corps in Kenya taught her that "it doesn't take much to push you into some kind of bias." She is fascinated by the ease with which one could assume a racialist mindset in a foreign place: "I got so I could tell people's ethnicity! They are all African peoples, and I became biased, culturally biased, by 'us' and 'them'! And I love the idea that it was so easy for that to happen to me, to get caught up in [a] set of cultural and social biases." Mountain said that even though she was aware of these growing biases—since she had already made an earlier, destabilizing passage from a more cosmopolitan England to the desegregating American South—there were "emotional things that actually had a huge impact" on her thinking about race which she couldn't shake. Referring to this mindset as a "tremendous -ism," Mountain expressed what a number of scientists also describe as a universal racialism. Scientists believe that only by acknowledging their own racism and racialism will the public begin to dismantle society's structural inequalities.

Mike Bamshad, at the University of Washington, also suggested that race should not be used as a proxy for genetic ancestry, but that there may yet be undefined factors of interest to the genomicist. A son of Iranian and Danish parents, Bamshad related several stories about medical projects he undertook that shaped his understanding of race. One involved a patient he saw on rounds who was African American and suffered from albinism. In case reports, the hospital referred to her as a "black albino." Bamshad subsequently wrote an article for his medical school journal on the discrepancy between color-coded descriptors like "black," social labels like "albino," and patient ancestry, and the inadequacy of all these taxonomies to describe the individual. In doing so, his suspicion of the commonplace practice of comparing "black and white" sedimented into a scientific framework, which he has carried into his development of population genomic methods. Similar stories populated other scientists' reflections on race, showing again and again that scientists draw on intricately woven personal genealogies to think through race and ancestry.

Bamshad discussed the importance of understanding other biological factors related to race: "Some would argue that there is something intrinsic to race that can't be measured that makes using the term appropriate. I think that that's fine, I have no problem with that, as long as they are explicit and that's

the reason that they're using it." He echoed the aforementioned call for genomicists to be clear with themselves and the public about their heuristic uses
of race. Among his many publications on race, he has written popular articles
for *Scientific American* and the *Journal of the American Medical Association*, and
has hosted a race and genomics issue for the *American Journal of Medical Genetics*. In these articles, he too petitions scientists to bring clarity to the public,
but to start with themselves.

Though all the scientists I met who privilege constructionist arguments
root their beliefs in their own past social experiences, scientists who came forth
as having a minority identity often spoke in terms of personal or familial suffering. One quarter of the scientists I interviewed responded to questions about
the meaning of race by describing their own experiences as a racial minority
in North America. These respondents display how scientists struggle to turn
negative experiences into socially beneficial working understandings of race,
and how their work is inextricably linked to their need to work through these
issues personally and publicly.

I met Georgia Dunston in her sixth-floor office at Howard University's
Cancer Hospital. When relating her understanding of race, she argued from
biological and spiritual grounds: "God did not make us all different races; I
mean, my science and biology says that we are all one human race with tremendous exquisite and beautiful diversity and variation in how we manifest that
race." Dunston's rationale defers to a broader cosmological view of life on earth
as a spectrum of genetic variation. Like many others, notably the current NIH
director, Francis Collins, who is known to the public as a born-again Christian,
Dunston doesn't attempt to hide the moral valence of her views.

Dunston told me that her childhood in the segregated South compelled
her to investigate what she called "the basis of the differences that we see in
what has been referred to as race." Early on, she understood that racial differences were responsible for social inequality, but she internalized the dominant
racist belief that African Americans were inherently bad and thus deserving of
second-class citizenship. She described her earliest scientific mission as a struggle to understand and rectify the innate errors of African American biology:

> It seemed if we could understand what made us different, then maybe we could
> be all alike. So that the problems—especially the negative things with folks who
> were in groups that seemed to be bad (as we were told)—if we understand what
> made us this way, maybe we can change them. . . . Like the society would say
> that this is the right way to be, this is the best to be, this is the ideal way to be.

Dunston analogized this interest in racial problem-solving to the palliative paradigm that frames biomedical genetics: "So, I guess, it's no different from if you define something as a disease. We wanted to understand the basis of the disease, so we could correct the disease and have health." Like other scientists I interviewed, minority and otherwise identified, Dunston explained the segregation-born racialist mentality in terms of a cognitive cage, from which she had not been able to escape until much later in life.

While Dunston and her Southern-born colleagues no longer take dominant racial discourses at face value—Dunston herself is quick to point out that she has replaced mentions of race with the phrase "[groups] that we call races"—statements like Dunston's show how the field's challenges, from its inquiries into race and human variation, compel scientists to think through these long-held ethical questions. She went on to describe for me her entry into human variation studies:

> I was looking at what we have called racial groups. And in America, of course, that was largely your physical features. So, we were looking at what's the basis for the skin color differences, what's the basis for the anatomical structural differences, where one side was right and what should be desired, and another was wrong and bad. *So, if I understood what was the basis for the difference, then maybe I could help to make it all right.*

Dunston periodically used this same phrase, "to make it all right," to express how her earliest views had shaped her subsequent values and strategies. Calling race the "very basis for the National Human Genome Center at Howard University," Dunston illustrated how scientists attribute a central role to early socialization experiences, and how they struggle to keep racial justice a top priority.

Thinking through the social injustices of race led many scientists who identify as minorities to stress their own rare existence in the field. In his former office at the University of Chicago's Medical Center, Rick Kittles talked about how his daily experience as a minority compels him to weigh in on the public debate over the biological status of race:

> I'm a black geneticist, and so there's this intersection of genetics and intersection of social-political race that I experience daily. So of course I'm going to be involved in this discussion. . . . These administrators and [representatives] of funding agencies . . . say things like, "Wow, you're the only one I know that can talk about some of these things!"

Kittles's comments show that minority scientists are hailed into racial deliberation by the very institutions they belong to and by the field's racial distribution. Kittles and the other minority scientists I interviewed welcome the challenge. They argue that their racial perspective affords them a unique standpoint from which to "magnify" the discussion and imbue it with alternative interpretations.

These scientists assess the field's progress from the perspective of minority justice, or how genomics is affecting minority well-being and broader concepts of race. As a result, many have begun questioning their own role in the development of technologies that could be used for racist purposes. As I listened on, Kittles repined: "I'm at the point now in my career where what is bothering me is I've contributed some methodologies and technology that's being utilized in areas that I didn't want to utilize." Particularly offensive to African American scientists is the forensic use of ancestry estimation technologies, which scientists fear are being applied in racial profiling by law enforcement. Minority scientists also criticize "the reification of distinction" that principal components analysis and other genomic technologies are being used for in population studies. Though these scientists equally rely on continental comparisons in their research—as Kittles argued, "We use these ancestry informative markers to say something about shared ancestors; this is continental ancestry, *and this is not race*"—they aim to vigilantly protect a clinal, or fluid, notion of human variation.[22]

At Howard's National Human Genome Center, Charles Rotimi elaborated the parallel fluidity of identity processes that genomicists attribute to the social construction of race. Married to an African American woman, Rotimi views her racial experiences and the racial experiences of their family as socially constructed by America's legacy of slavery, whereas his own socialization was of a different nature. His experiences of code-switching between his primary ethnic taxonomy, his current racialized identity in the United States, and emerging ethnic and racial classifications in Africa have led him to perceive racial identity as fluid. Rotimi described his work and that of the Genome Center as an effort to use "genomic knowledge to inform our *whole* issue of identity" so that "with more understanding of human genetic variation we will begin to see that these boundaries that we have created are really not fixed." Representing a stance I find common among scientists who have been racialized into a minority status in the North American context, he maintains that the genomicist's goal is to replace the particularist notion of race with a universalist notion of "human genetic variation."

Rotimi called attention to the scientist's power to represent diversity in politically charged ways. He told me that when discussing human populations he prefers to refer to continental populations as "historically separated," as opposed to "distinct." Racial identity is thus, in his estimation, a thing apart from genetic clusters, evolution, and the distribution of biological variation. This nominal shift to the conceptual language used by scientists is not unconscious; for many, every act of communication holds a potential to rectify misconceptions and an opportunity to shift the terms of public understanding.

When conceptualizing race in a constructionist idiom, some scientists present deep ambivalence about *any* existence of groups. Eric Lander sat back on the sofa in his office at the Broad and helped me to visualize genetic variation in humans:

> Draw a big graph of six billion people. For each person you have ten million common genetic variations within the population. Draw a line between person one and person two for the first genetic variation if it is the same, and a line of a different color for the second genetic variation if it is the same. Six billion nodes on the graph. Between any two nodes, I have ten million lines, which I'm going to color in with the color if they are the same. That's the total description of genetic variation in a population. That's it.
>
> Even if there are more connections in *this* region of the graph, or in *this* region, it doesn't mean I can take my magic marker and draw a sharp edge and say everybody in this region is Asian. Because I'm sure, as I move across the Asian continent into the European continent, there are gradations such that there is no short boundary.
>
> So does that mean race exists or does not exist? If by race existing we require a rigid sharp boundary, well it is problematic. It does not exist. If, on the other hand, you mean that there are regions of this graph that are somewhat more connected than other regions, then it does.

Lander's point is that in the vast domain of biological difference, variation is a multilayered spectrum of relatedness. Decisions about which relations are important are, scientifically, arbitrary.

Of course, Lander is fully aware that those decisions are politically meaningful. Yet he wants scientists to consider the logical end of categorical models—"By a category do you want to mean the ability to assign an identity card to people, to every person being clearly identified?" Lander is critical of the media for reducing the vast complexity of the human genome to race, and making it appear that genomics supports simple answers. He prefers an open model that will preclude the production of race-based medicine, continent-specific therapies, and any population-specific applications.

At a pub not far from his country home in rural Virginia, National Geographic's Spencer Wells told me that race would become obsolete once people had all their genetic information on hand. He acknowledged that the body's superficial phenotype is important to laypeople but said that disease phenotype is the concern of genomics. When Wells expressed exasperation that race continues to loom large in the public and in genomic discourse, I asked him what he would like people in his field to do instead. He answered:

> I guess to focus more on the specifics of ethnicity and to use that as a tool to start to investigate some of these correlations, genetic correlations. Race is a very blunt instrument, genetically speaking, and it is not terribly helpful. Speaking as somebody who has done a lot of work in Africa, there is just amazing stuff that comes out every time we dip into the African gene pool. There is stuff that you never expected to find, and it is a reflection of that long history in Africa. If we take the mitochondrial Eve 200,000 years ago and that time of the first appearance of humans in the fossil record as the dawn of our species, we have spent three-quarters of our history as a species living only in Africa. And so there is a lot of stuff we still do not understand about that continent. To group all Africans together into a single ethnicity or race or population group, I think, is ludicrous. So I'd like to see people be a little bit more specific while they are looking for these actual correlates. And ultimately, that is what they want to find.

Again Wells referred to the African gene pool as simultaneously exceptional and exemplary of the intrinsic diversity of the groups we consider races. Wells holds up his Genographic Project as an example for other scientists in the field. Though this project is meant to be anthropological, he insists that proxies can also be eliminated from the medical field. Wells plans to shape the field of personalized medicine to this extent from his position as a science advisor to Navigenics, one of the leading personal genomics companies that provides medical as well as genealogical counseling to its clients.

Lander, Wells, and the other scientists I met are not advocating for genomicists to ignore race; rather, they value dialogue about its social construction. Wells told me that racial education had been a part of his earliest training in population genetics, especially while studying with Cavalli-Sforza. In Cavalli-Sforza's lab at Stanford, interrogating race and its relationship to patterns in genetic clustering was scheduled into the weekly lab meeting. In Lewontin's lab at Harvard, Wells continued to engage in ongoing conversations on race and human variation, which were built into lab life. Recalling the centrality of race in his education, Wells spoke of a commitment to keep it that way through his own public engagement for the National Geographic Society. Thus, stepping away from a genetic concept of race does not mean that genomics relinquishes its participation in racial discourse. On the contrary, it means an ongoing deliberation over race.

Finally, in offering constructionist explanations of race, scientists also advance models that defy nearly all forms of human classification. Amid the din of clinking glasses and blaring music at New York's Mercer Hotel, Craig Venter firmly stated, "I say [race] is a social concept, not a scientific one." He added, "I don't think ancestry's right either." This leading innovator explained to me why he has little faith in the field's characterization of human variation. He argued that the field is preoccupied with "ethnogeographic backgrounds"— geographic origins that connote similar evolutionary responses—and any signs of "inbreeding," but that this is an imprecise measure of meaningful difference. Venter criticized the ethnicity-matching protocol of case-control genomic studies and explained that diploid genomics, the sequencing of an individual's entire genome, is the only way to begin to classify humans in any meaningful way: "There are reasons to classify people. It's worth looking in at people that have common biochemical origins for their risks for hypertension or for prostate cancer or colon cancer. Any one person can be in all three of those groups or none of those groups and all kinds of others." Venter called race-based medicine his "biggest pet peeve of all." "Everybody talks about people who have recent African descent or African Americans that have higher incidents of prostate cancer. Yes, but that does not mean that therefore you look at every person with black skin and say they have a greater chance of prostate cancer!" Venter claimed that race is a ridiculous concept because there are "no purebred humans of any type," only "people with longer lineages." He said that even with first-cousin marriages in isolated places, one only has to go back a few generations to find exogamous behavior. In other words, he sees every living group as interrelated and admixed.

At Johns Hopkins, Aravinda Chakravarti similarly said that race makes him "queasy" because it doesn't perform the way he needs it to. He is particularly known in the genomics community for his belief in grid sampling, or laying a grid over the globe and sampling DNA at equidistant coordinates. Grid sampling is an expensive process that permits no prior notion of ancestry to guide computer programs in assessing genetic similarity. Chakravarti called "the population in population genetics" a "synthetic concept." He further explained: "I would first say that, yes, different human groups are different. But human groups don't show the kind of discontinuities as we expect the word 'race' to have. . . . It's a problem we need to solve. It still happens, and people still talk of white and black and pink and green and all kinds of races." Chakravarti continued with an example from genetic screening: "It's easier to test every newborn and every newborn that's thought to have [phenylketonuria], put them on the special diet, and it prevents them from becoming mentally retarded. So your group affiliation is *unimportant*. I am very strong about this: it's *unimportant*." Counterposing the color-coded system of race with life-and-death matters, he ridiculed the use of social identity to make medical decisions. Like Venter, he feels that scientists should consider only the genomic data present today.

Still Chakravarti did not shy away from discussing the social construction of race. He spoke at length of his role as a consciousness-raiser in the field. Part of that stems from his feeling that the average American lives in a "poverty of difference" and is thus inexperienced in social diversity, especially the cosmopolitan diversity Chakravarti has experienced in other countries. He told me how his multicultural upbringing in Calcutta spawned his interest in population studies:

> I grew up in India (in Calcutta), and so I grew up in a country that had at least three or four histories and in which people—Calcutta is a large city which had an older history, and then there was the history of the Moguls who came and the history of the English, and that was the first capital of India, and then there was a modern political history . . . and it's a place where, when I grew up, there were people who looked almost as different as what one would say the many races of human beings are. So, I had an early interest in genetics and I think this was cemented.

Chakravarti went to a small college that had a history of using Calcutta as a living lab for colonial population studies. There he tackled the troubled history of population science with undergraduates of all backgrounds: "The motto of this institute was 'Unity in Diversity,' and I think that's pretty much a motto of where I grew up. . . . People say this was a third-world country or whatever, but my friends in Calcutta were Hindus, Muslims, Christians of more than one kind, Jews, and Parsi." Chakravarti sees ethnoreligious diversity as particularly salient and complains that his social life in the United States was largely limited to Christians and Hindus. As he mulled over America's "poverty of difference," he stated that because of his upbringing, "it would be impossible for me *not* to be interested in what I'm interested in. In fact, I'm surprised that the rest of my coworkers weren't interested in what I was interested in!" Like other foreign-born scientists I spoke with, Chakravarti translates his experiences into a starkly particularistic viewpoint. During his presidency at the American Society for Human Genetics, Chakravarti often critiqued the notion of racial biology.[23] He is now working on a book that will draw on genomic scholarship discounting the existence of race; here he and his colleagues can further explore the group concept and its implications for racial identities and ties.

While these perspectives on race contrast the earlier biologically deterministic definitions insofar as they deny the existence of race, they too advance a complex way of understanding race that paradoxically has the potential to bring society back to a biologically determined definition. On the one hand, constructionist viewpoints suggest that genomicists must adhere to a purely social model of race; on the other, there is no guarantee that the genomic classification system they prefer would not function similarly to typological race. Despite this conundrum, the ethical standpoint that pairs with such a constructionist model rings synonymous with the prodiversity, culturally sensitive themes proffered by determinists and weak-correspondence advocates. As with the scientists who use racial categories to target minorities, these scientists produce community interventions that strategically use race in order to generate inclusion and participation.

From Concept to Practice

In the field's predominant understandings of race we see that most models are pliant enough to lack a rigid biological foundation, yet they uphold a continental framework and meld with common lay notions of race. I see this as antiracist racialism: a confluence of models that perpetuate the study of populations as discrete entities, though motivated by the common drive to create equity between them. Scientists shuttle between deterministic and constructionist positions as they think through the practical and ethical dilemmas that are part and parcel of their job.

The articulations of race we have here explored show that genomicists adhere to a range of socially and biologically inflected beliefs about race and ancestry, each attuned to a set of political and ethical values that scientists openly promote. From stark genetic notions of race that conform to historical typological characterizations, to explanations that emphasize identity processes and even the scientist's ability to present genetic populations purposively, genomicists employ models that can serve minority health and promote racial equality. Contrary to many critics' assumptions that scientists don't "have it clear" for themselves and don't care about the social and political effects of their findings, these scientists are attempting to square their values about racial justice with their interpretations of human difference. They make sense of the large-scale minority studies or research programs they lead by considering their personal experiences and the future they want to produce for themselves and others they care about. Their thought processes show the personal in the political, and the political in the personal.

It is precisely because genomicists hold heterogeneous concepts that race remains of interest to the field. Scientists who offer genetically or biologically deterministic explanations for race want to use genomic tools to reduce racial health disparities. But similarly, scientists who draw a line between race and ancestry believe genomics should incorporate more social epidemiological tools with which to study racial health disparities. The division here is actually between those who would use *one* classification system for race and ancestry and those who would use *two* separate taxonomies. Genomicists are thus quite unified in several assumptions about race: genomics has an influential role to play in solving racial dilemmas; political and ethical values are fundamental to genomic inquiry; and scientists should bring their personal expertise into their research.

The values that scientists share produce a moral order intent on finding ways to bring minorities into the genomic fold. It is necessary then to look closer at the way scientists operationalize these definitions and concepts. What strategies have genomic elites devised to ensure that their values will be put into practice? Exploring the fieldwide sampling and reporting strategies that often go unexplained in genomic research, we can begin to understand how individual scientists carve out socially conscious ways to work with race by negotiating the practical demands of research and their own normative frameworks.

5   Everyday Race-Positive

THE POLICY MANDATE to actively include ethnic and racial minorities in all scientific research has been increasingly incorporated
as a standard for research involving human subjects in the United States.[1] Accordingly, communicating and thinking in a multiracial, race-positive idiom
has become a norm for the field. But how do individual scientists put these
broader norms into everyday practice in their labs and projects? How do their
personal moral convictions shape the ways in which they construct and use
sample taxonomies?

One inclusionary campaign stemming from one of the first whole-genome
analyses of African American and Latino men provides interesting answers to
these questions. In 2008, Life Technologies' Francisco De La Vega and Stanford
University's Carlos Bustamante set out to study patterns of admixture in two
underrepresented American groups with the hope that their research would
usher in a new focus on "the contribution of native American genetic variants
to the disease burden in the Americas of today."[2] Bustamante, a Venezuelan-
born recipient of a MacArthur "genius" award, rebuked the field saying, "One
of the reasons that researchers say they study White populations is that they're
easier to study, they're more homogeneous, blah-blah-blah. . . . But, it's really
that they haven't really done enough to engage minority populations."[3] From
2009 to 2010, these researchers partnered with other California scientists to successfully petition the 1000 Genomes Project to include the DNA of five hundred African Americans from the southwestern and southeastern United States,

Afro-Caribbeans from Barbados, Mexicans from Los Angeles, Peruvians from Lima, Colombians from Medellín, and Puerto Ricans from Puerto Rico.[4] Still interested in expanding their database beyond the European, African, and Asian samples sourced from the HapMap Project, the National Human Genome Research Institute seized the opportunity to bring Bustamante on to find key scientific liaisons needed for work in South and Central America. Back in the United States, Bustamante championed the work of minority-focused scientists like Esteban Burchard and Rick Kittles, arguing for greater funds to be devoted to the institutionalization of their research.[5] These efforts led to an enlargement of the institute's database and an enrichment of its racial paradigm. Sponsoring individual genomicists to deploy their identities as minorities, the field is making race-positive science a reality for these scientists and the groups for which they feel socially responsible. The ultimate goal is to develop drugs and biotechnologies that will be more sensitive to these groups' biological needs.

Issues about the racial identification of researchers and research subjects are important because in genomics human taxonomy and the question of whether to use race are always elementary concerns. All human genomic research begins with the definition of research populations and decisions about how those populations will be addressed. All studies produce labels for samples and research populations that are later reported in the field as well as to the public. Beyond these basic steps, all studies connect to a body of research that is dedicated to mapping the world's human diversity and defining the nature of human difference. Genomic taxonomies thus have a significance that extends far beyond the lab.

Just as scientists make sense of race by drawing on their personal experiences, so do they manage the practical landscape. This means pragmatically drawing on common lay understandings and government classifications while infusing their research with personalized political investments. As the example above shows, genomic research is beholden to identification processes that affect all members of a racialized society. Scientists' identity politics are played out recursively in the choices they make for sampling, recruitment, labeling, and reporting.

Because elite genomicists are committed to expanding their purview to research the social factors that contribute to health disparities, to build a better relationship with the public over the biology of race, and to square their own values with their scientific practice, most also continue to sample in ways that are amenable to drawing conclusions about race.[6] My analysis demonstrates that taxonomies are individually imagined but also collectively fashioned, and they simultaneously pull from and resist broader social meanings about whom to include in research, how to respectfully access minority populations, and what techniques and resources to garner to do so.

Formulating taxonomy is a political act that emerges in a social context in which race is a defining principle of difference and shapes the go-to taxonomy for most labs. As sociologists of knowledge and race have shown, taxonomies are forged between the demands of everyday life and a store of hegemonic ideas, such as the substratum of racial and gender taxonomies.[7] In everyday life situations, ideas and conventions form relationships that successively enforce and strengthen reigning assumptions.

Designing Sample Populations

Until the inclusionary turn in genomics, scientists considered endogamous human breeding groups equal to the races of the animal world. Sampling humans was no different from sampling earthworms or fruit flies. At the same time, sampling human DNA was done only in the case of research on Mendelian, or single-gene, disorders that traveled in families. Scientists genotyped related family members in the search for potential causal variants. Population diversity in humans, especially on a global scale, was not measured through interrelated genes but rather through blood types and comparisons of proteins. Yet with the development of DNA sequencing, fragmentation, and amplification technologies, scientists began moving toward genome-wide analysis and rethinking racial population comparisons.

Though the major international genome projects like the Human Genome Diversity Project and the International HapMap Project have established large databases of samples, individual studies often need to create their own sample bases to answer questions about populations not included in those databases or to understand genomics based on different technological criteria. For example, Neil Risch, Hua Tang, and Esteban Burchard at the University of California and Stanford University have collected the DNA of African Americans and Puerto Ricans in order to understand admixture, while Ken Kidd and Sarah Tishkoff of Yale and the University of Pennsylvania have culled DNA of Biaka of the Central African Republic to perform haplotype mapping studies.

The first step in designing a genomic study is choosing which populations to sample. Remarks by Joel Hirschhorn sum up how genome-wide association investigators control for ancestral "background noise" in cases and controls: "[A study] might not work because even though there might be a group, there might be a correlation with ancestry that might not be due to underlying genetic differences." In this instance, by "genetic differences" Hirschhorn refers to the disease-causing factors of interest, and by "ancestry" he refers to something akin to race or ethnicity. In Hirschhorn's GIANT Consortium studies, he draws samples from Bavarian and Finnish biobanks among other sources.[8] Like all genomicists, Hirschhorn attempts to match the race or ethnicity of the study population with those of his control population. At times, genomicists will also consider a range of other social variables like language, culture, political and national identities, and geographic proximity, which they try to keep constant between the control and study groups. Scientists then statistically estimate the genomic ancestry of the samples with ancestry estimation technology to make sure that subjects' genetic profiles match their self-reported identity in the study's terms, and are thus appropriate to use in the study. All this means that scientists start with social classification systems and end up either confirming or disconfirming the individual DNA sample's place in those systems.

Scientists manage the selection of research populations in part by what they know about how the wider public uses social categories, and what political connotations they might have. As Noah Rosenberg explained, though scientists control for social factors by estimating the genetic ancestry of their samples, "the genetics work is not completely divorced from social categories, even if the results of genetic clusters can be obtained without using previous information." Joanna Mountain put it this way: "The [categories] don't come from biology; they come from social history." Though these scientists are aware of the social origins and political valences of the categories they use, these categories nevertheless build a framework from which they and their associates perceive and explain biological variation. This points to a common pattern in the construction of racial knowledge termed by Sarah Daynes and Orville Lee as "the dialectic of belief."[9] Genomicists utilize their memories of prior racial experiences to interpret and produce meaning about the natural world. Although there is an explicit physical reality that is their task to represent, members of a deeply racialized society like the United States are necessarily interpellated in politicized debates. Acts of articulating racial variation are thus "racial practices"[10]—the actualization of beliefs about prejudicial treatment, racism, and racial inequality in everyday life.

Genomicists begin by listing potential sample populations in terms that help to interface with the public, even if they are unsure how those terms will play out medically or genetically. This means that the lead investigator's social and political awareness and sensitivities play a large role in determining whose DNA will be sought for a study and how populations will ultimately be included, stratified, and compared. The questions that leading scientists pose depend on whether they are familiar with a given social group and take that group's medical or cultural concerns seriously. Researchers who primarily work on hybrid public-private projects, such as the U.S. Cancer Genome Project[11] or the 1000 Genomes Project, also face pressure from academic and nonprofit institutional hosts to interpret health and evolution in ways that are relevant to the major social groups in the national population. Accordingly, Census race typically serves as a baseline, default taxonomy.

Not all studies compare variation across major human populations. There are studies that home in on regional population diversity and those that compare DNA on a global scale. By way of example, David Reich described his 2007 study on population differences in Luo, Yoruba, Masai, and Kikuyu[12]: "We went to areas that were ethnically one group and we required people to have all their grandparents from there living in the same area, and to say that they were from the same group, or report that their grandparents were all in the same group. Often, we say: 'If your grandparents self-described as . . .'" Grandparental self-reported ethnicity is considered to be a best proxy for shared ancestry, since it is assumed that people two generations back were less likely to relocate from their ancestral homes.[13]

With *globally* comparative studies like the Multiethnic Cohort study, scientists give research subjects a limited cluster of choices for self-identification.[14] Reich continued:

> We ask them: "What is your self-described ancestry?" And I have a list of things that they can fill out. They will say, "European American," "African American," "Latino American," "Japanese American," etc. That is what I usually mean when we use self-described ancestry. Then, you can use that to do the genetics on people and see whether those self-described ancestry categories correspond to anything genetically.

These descriptors follow a continental schema that in Reich's case are designed to avoid connotation with color-coded racial terms but correspond to categories the public can easily identify with. These are terms that are designed to get at the "real" genetic ancestry underpinning the major axes of social difference yet in ways that are still socially recognizable.

Scientists who lead sampling expeditions form a personalized relationship to the sample design process. In their experiences collecting DNA samples themselves or directing a team of researchers in sampling, scientists attest that unless they are taking part in a consortium project or are being funded for research on a specific population, *they* choose whom to sample. This involves building ongoing relationships with certain social groups and striving for consistency with favored population descriptors. For example, Burchard works with Latinos from the San Francisco Bay Area, while Mountain has conducted ongoing research with ethnolinguistic groups in Tanzania.[15]

Yet scientists also share samples that are preidentified by others in the field. Lynn Jorde illustrated how he amassed the sample populations that he has famously used in his lab's global cluster studies:

> Some of them we did. Some of them were shared by others. We have some samples from southern Africa that were originally collected by Trevor Jenkins. In fact, one of my old colleagues, Henry Harpending, was holding a lot of that collection way back in the late sixties. So I connected with Henry and said, "if you could, get DNA samples from Trevor, because we really can't afford to go get these populations," and Henry said, "Well, sure I collected a lot of them myself" . . . that was about sixteen to seventeen years ago. . . . We got a few samples from Coriell, so we have some samples from Pygmies and a few lines from Asia that came from Coriell. Ken Kidd gave us some cell lines. And then we had a colleague in pediatrics who was studying Vitamin D rickets in Africa. And when he was there, he collected quite a large number of Pygmies, some Nilotic-speakers, and Bantu-speakers in the same area—what was then Zaire. So that's where we got probably another hundred or so samples is through Bill Fischer (he's a coauthor on some of our papers). So almost everyone who has been putting together data the way we have has done the same thing: they get samples when and where they can, because it's hard.

Jorde and other scientists explained that for reasons of cost and efficiency, genomicists are selective about when to launch a sampling expedition. Sampling expeditions are capital-intensive, and most require funding from the scientist's host institute, government, private foundations, and industrial sources. Furthermore, today's amplification technologies do not require as much DNA as they did in the past.[16] That means researchers can share small amounts of DNA. One could say a little sample goes a long way.

At the same time, this means that old and new sampling frameworks converge through sample circulation. Jonathan Pritchard explained that samples already in circulation can always be reassessed for genetic similarity using new genomic software. Also, as Jorde's comments show, taxonomies from prior expeditions are inherited and supplemented with newer material and then passed on to other labs. In this instance, Jorde's collection includes samples collected in the 1960s, 1990s, and in recent years. Many of those expeditions were dedicated to gathering data for studies of evolutionary divergence of continental races. Thus, the taxonomic assumptions built into these postwar projects form an inconspicuous scaffold for today's research.

Reflexivity and Reciprocation in Recruitment

Scientists actively take steps to extricate their work from the past paradigm of colorblindness and make sure that the new paradigm of sociogenomic inclusion governs their sampling methods. In fact, the elite scientists with whom I spoke use past paradigms as a foil for building a new paradigm. They are aware of the ways that evolutionary synthesis scientists engaged the colorblind antiracist paradigm of the postwar era,[17] which ended up producing racial maps of human diversity and reinforcing assumptions of racial difference. Today's genomic scientists, in response, carefully craft sampling taxonomies that critically interrogate variables formerly taken for granted. For many, this leads to a particular type of identification with, and embodiment of, their research habits.

Charles Rotimi's critique of genetic epidemiology evinces this nicely:

> When I look back and I really look at the way we were taught and socialized in terms of group differences, you really have to say that it was really, really done very poorly.
>
> I went to one of the best schools that I can easily say that taught the students who are now the best people in epidemiology. But, in general, the whole concept of having a variable for race coded as 01 or 04 or H1 for black! And then use that as a way to sort of capture people's whole experience? Apart from those other things that you have said that they have measured like how heavy they are, maybe income, or educational level? But forget it.
>
> So in a sense, what people are doing is sort of saying, "We are trying to understand A. We know that A = B + C + E + through Z. Right now, I have tools to do this. I can't measure this or that thing, but I have another variable here called race. That sort of gives me an idea of what all of these other things are, so I am just going to dump it in here.

Rotimi is not content to use race as a proxy for all sorts of unmeasured variables. He prefers to go out and measure those criteria himself. He and many others describe their job as creating a new sampling paradigm marked by greater attention to the complexities of racial health disparities and interrogation of the race concept itself. At the Center for Research on Genomics and Global Health, Rotimi distinguishes between African Americans, Afro-Caribbeans, and West Africans.[18] These scientists are upset at what many have referred to as "laziness" in matters of classification, and they seek to extricate genomics from uncritical taxonomic science.

As genomicists critique past paradigms, they emphasize the importance of giving back to sampled populations. Remarks by Rick Kittles illustrate the value that scientists place on reciprocation:

> Back in the days this was called helicopter genetics—where you swoop in and get the blood samples and swoop out, and you never see [the research subject] or talk to him again. That's another example of that thing where you know the drug company says, "Here's the drug. We think it works. Let us swoop in and get some of these patients and see if it works and then market it to them."

Kittles draws our attention to the exploitation that has historically pervaded foreign sampling. Like many of his colleagues, he argues that scientists should bring a more personal dimension to their research to ensure that individual subjects' needs are equally considered. These scientists also join critics in viewing prior genetic research as "salvage anthropology," a phrase genomicists use to denote the common practice of flying into poverty-stricken regions of the globe only to take DNA for Western academic purposes.[19] As with the major projects, individual researchers hold dialogues with community leaders to determine what kinds of health resources can be exchanged for participation in studies. Kittles has focused on relations with West African groups like the Kpelle and Bubi and African Americans from Chicago's South Side.

Similarly, scientists cultivate what they view as cultural competency. Rotimi, for example, chastised projects in which ignorance of particular cultural practices prevents minority health intervention:

> But there was an example of a new study going on, in one of the universities like Hopkins or something like that, where they did not really have good access to the African American community. And over about a year or two years periodically they only called in one or two families.
>
> And somebody says, "We cannot or we do not want to participate in the study"—but [you are] forgetting that the reason maybe you are not being successful is that you do not understand this community. You don't have some-

body that they can relate to to participate in a study like these. And of course you may also have a history of collecting this kind of information and never going back to them with your results. And so why should they come again?

Scientists are dismayed by the existence of urban biomedical research institutes that ignore local minority communities. Many demarcate a boundary within the field between culturally sensitive and insensitive genomics. They are greatly concerned that some institutes are giving the whole field a bad reputation with minorities. Their position reminds us that ethical stances and the pragmatics of research are interwoven. The large-scale epidemiological projects and for-profit enterprises led by genomicists like Rotimi and Kittles could not survive without alterations to sampling practices that produce bonds of trust between researchers and the communities that subjects belong to.

As a result, scientists are fostering minority-conducted sampling expeditions. Rotimi's own path from founder of Howard's Africa America Diabetes Mellitus study to today's Human Heredity and Health in Africa has compelled him in recent years to take a leadership role in developing a model ethical framework for the field: "For example, when I am in Nigeria and I am working among the Yoruba, I hire Yoruba people so that we get fewer consequences. I do not go there and start bringing people from Kenya to come back and do the study, no!" Here as in the identity politics of recent decades—and in ways reminiscent of past colonial policies of indirect rule—it is assumed that recruiters with the same ethnic background will be more sensitive and thus successful. Often ethnic groups that are recruited are made to stand in for a continental population; thus, the racial identity of recruiters is equally at stake. Because genomics puts a high price on insider access to vulnerable populations, it holds a growing commitment to having minority scientists operating at the frontlines. Rotimi's comments convey the inextricable links between sampling paradigm, access to samples, and downstream political consequences.

A number of scientists draw on social scientific streams of expertise—especially around anthropological, bioethical, or historical training—to formulate their reflexive approaches. Kittles referred to some colleagues as too "hardcore geneticist" or "card-carrying geneticist" to see their work as anything but "genotypes and alleles."[20] He applies his formal training in biological and cultural anthropology to "interpret and design a little differently." Kittles added:

> [Some researchers] don't know how to operate. They don't know how to talk to people. They don't know how to build relationships. They don't know about trust. They don't know about respect.

When I go into West Africa, people see themselves in what I do, and I make sure of that. I'm very sensitive, even in the studies that we do in the African American populations. So I think that there is a level of fraternity and actual egos that create barriers to sample collection.

Kittles explained that the groups he samples in Western Africa are enthusiastic to give DNA for biomedical and anthropological purposes, "because they think that it is going to bring them closer to African Americans and then with that closeness comes support, lobbying, and tourism." "There are people who actually go back after they get tested, and there has been a lot of lobbying for certain communities. There have been requests for dual citizenship! Like the Cameroon Embassy has been hit *hard* for dual citizenship. The Nigerian Embassy also. And I think there's a lot of travel back and forth." Identity politics, refracted through the economics of sampling and citizenship, frames these endeavors to include and represent domestic and foreign groups. Scientists open channels of aid and resources in order to gain access to DNA that is relevant to the study and marketing of local lineages, including that which is believed to shape the health of U.S. racial minorities.

Scientists envision themselves as bringing together good science and good cultural politics. Many see themselves as the stewards of a new humanity. Their efforts loop back into the framework of "living *as*" described in the previous chapter, wherein scientists collectively sponsor racial identifications that conform to a set of behaviors that are expected of membership in particular groups.[21] This marks whose health they should address and whom they should serve, but also whom they should rely on and seek recognition from. Living *as* becomes a natural framework within the halls of science because it governs scientists' identification processes in the greater social sphere, such as in familial and casual settings outside the lab.

The end result of this is that scientists use their cultural competency: to make lasting ties with the communities they need for their research, to expand their databases into the developing world, and to create richer maps of global diversity based on an understanding of race forged in their personal lives. Trust is both a value frame and a key to access to social groups.[22] In the case of Kittles's database, his samples serve as the reference populations for part of the field's African diaspora biomedical research, evolutionary research on African ancestry, admixture mapping, and his private company's genealogy services. Just as African Ancestry's public reputation is built on the visible positive relations between reference groups and clients, the wider field of genomics enjoys a more favorable public status and greater access to vulnerable populations be-cause of its values-based sampling paradigm.

Targeting Minorities

Genomic elites use a host of racial practices to ramp up their social justice efforts. Scientists fervently agree with the federal government and the major genome institutes and consortia they take part in that prior research has been Eurocentric. They explain the Eurocentric sampling bias demographically—as Pui Yan Kwok put it, "we concentrate on people of European origin in this country because mostly that is what [is here]"—but scientists also feel that re-searcher interests play a role. Luca Cavalli-Sforza used Jews as an example of an overrepresented group whose interests have dominated genetics. Cavalli-Sforza reasoned that Ashkenazi Jews have been particularly overstudied because of their history of participation in the medical field. He complained that most populations "are not as well-known, and we don't know what diseases they have." He claimed it has been the fault of genetic researchers; thus, it's up to the latest scientists in the field to change the pattern.

Nearly all the scientists I met shared their protocols with me, which they have designed to achieve inclusionary goals. An overarching strategy concerns oversampling minority populations. In the sleek Silicon Valley headquarters of Perlegen, associate director of biostatistics David Hinds provided the figures for Kwok's conjecture:

> So, if you just sample people at random in the United States, you've got 7 per-cent Asians and 9 percent African Americans and 6 percent Hispanic people. But with those groups you are not going to have enough of them to be able to detect small genetic effects, and most of your analyses are going to end up being restricted to the subset of European people. European descent—that's the single biggest group you have there.

Perlegen's multipopulation studies, such as its publication of the haplotypic sequence of chromosome 21,[23] show that the company's database has been built by oversampling in the vein of Census race. For these scientists it's not enough to "want" racial equality with an idea that one should be colorblind, because in a racialized society this passivity ends up further supporting those already in power. In the new paradigm, people have to come up with new strategies and methodologies that proactively subvert the status quo. Part of this process requires them to reassert racial differences. Perlegen is the company that made the largest private sequencing contribution to the International HapMap Project, and as shown in Chapter 2, it conducted its own SNP-mapping study alongside these efforts. The oversampling tactics that are produced in such an environment illuminate how mesolevel institutional norms intersect with broader cultural norms and individual research objectives.

Hinds explained that industrial population studies always begin with broad continental categories by which minority groups can be oversampled, and thus evenly represented, because it is too costly to do otherwise: "We're still at the point where the cost of the experiment for each individual person is high enough that you don't want to be doing experiments on a lot of people if you are not going to be using their data." Private companies amass reference population biobanks that research institutes and other pharmaceutical and biotech companies trill for genome-wide association studies. Later these population-specific findings can be marketed to the very social groups that researchers started with. In the case of the HapMap Project, Perlegen sampled by Hap-Map's Europe-Africa-Asia taxonomy and published global variation findings about the different prevalence of haplotypes in those groups; subsequently, it had an investment in producing therapies with that taxonomy.[24]

Genomicists also engineer inclusion by creating drug studies that target minority communities. Tony Frudakis described his research as "evolving smart drugs like GPS-guided weapons against disease, rather than dirty bombs." Looking at genetic ancestry in terms of local adaptations to xenobiotics, these scientists form relationships with social groups that can stand in for a genetic population. Yet the social impetus to deliver care to underserved groups makes the genomicist's job one of appropriate recruitment for the end-stage market. This can bring about a drive to develop medical cures for racial groups, despite most researchers' opposition to race-based medicine. As Mark Daly was quick to point out, "downstream it's been with more of a focus to make sure the research we are doing is as inclusive as possible to genetic backgrounds—racial backgrounds—across the world." Though researchers might want to start with targeted social categories then measure ancestry and design therapies based only on the genetics, delivering results to minority groups forces scientists to return to a social taxonomy when constructing their target markets. In the process, sampled populations, genetic populations, and target markets are conflated.

To target minority health, genomicists must first invest in efficient tools for the job. The field's original microarrays and software were designed for research on European American samples, which are believed to be more closely related and thus cheaper to sequence. As directors proactively seek funds for new technology, they necessarily reflect on the field's uneven analytical terrain. Esteban Parra, an admixture expert at the University of Toronto, lamented the exclusion of minorities in small-scale sequencing. One of Canada's leading genomic researchers and a longtime scholar of population diversity in the Americas, Parra described a field where many individual labs can't afford to use the most efficient biotechnology, even as larger studies fail to capitalize on their power to do so:

> See, it's a little bit sad that 100 percent of the big association studies are using the fancy [computer] chips. Of course the whole-genome association approached using the [Affymetrix] chips or Illumina chips; you can see that all of the studies having been published—the four or five biggest studies analyzing thousand and thousands of samples—they all have been done in European individuals. I think that is also not fair in terms of fixing health disparities.

Parra's comments provide yet another example of the foundational assumption that Eurocentrism is bad business. They also exemplify the motivations behind some researchers' attempts to fundraise for disparities research. Parra's lab has teamed up with Mexico City's Medical Center Siglo XXI to purchase Affymetrix 5.0 microarrays for diabetes research in Mexicans and Native Americans.

Jorde ran through the cost calculations that an academic lab makes to get the preferable tools: "We use the 250K chip right now because we got a bunch of them for $125 a piece. They're not that expensive anymore. They tend to start off high and then, like computers, as soon as something better comes along, everything else becomes cheap." Jorde explained that his lab was preparing to work with Salt Lake City's biotech giant Sorenson Genomics. Sorenson had recently collected over eighty thousand new samples, and Jorde was excited that they would be able to use their more powerful microarrays for "$375 apiece."

Jorde illustrated how public-private collaborative studies facilitate inclusion. Academic researchers often rely on well-connected pharmaceutical or biotech companies for the difficult process of consent and enrollment:

> We don't have an [Institutional Review Board] every hundred miles, . . . just all kinds of issues that say, well, this part of the world is probably going to be undersampled for a while. That's one of the reasons why I'm excited about this collaboration with Sorenson—they just spent a few weeks in Mongolia on horseback getting three thousand samples!

Here Jorde alludes to the institutional and practical limitations that academics face but that private firms easily supersede. Academic scientists also form networks with well-financed governmental research bodies to manifest the big science projects that try to garner global inclusion.

To measure the relationship between racism and genetic ancestry, several scientists are tackling the most taboo of racial issues: the genomics of skin color. Parra told me that he targets "superficial traits" to unpack their genetic significance and show the world that "pigmentation is just one kind of type of the distribution of variation that you see." He believes that illuminating the "complex scenario" of pigmentation mutations in their evolutionary contexts will lead to an enlightened social understanding of race: "If you understand how evolution happens, at the genetic level, then it is difficult for you to just fall in to those broad statements and simplifications."

Parra and the other scientists who work on pigmentation deeply personalize this task. They said that they feel hurt and frustrated by scientific misrepresentation of the genomics of skin color. When talking about his research into the relationship between craniofacial structure, pigmentation, and ethnic identity, Mark Shriver explained that in trying to improve modern medicine and forensic science he feels personally misunderstood by the wider scientific community. Shriver has gone on television with his own admixture profile—showing that he has an unusually high percentage of African ancestry for someone who for all other purposes identifies as white—in order to make visible the incongruence between pigmentary phenotype and genotype.[25] He insisted that his biotech applications would find their way into the medical mainstream. Indeed, his work on the population genomics of skin pigmentation genes has now been replicated by some of the leading genomic research institutes in the world. Even former critics champion these projects as being cutting-edge.

Aravinda Chakravarti considered how far to take genomic exploration of variants associated in the popular mind with race:

> So should we do this to study, for example, pigmentary differences between different individuals? But some have come from genes!
>
> The question is, what does [pigmentary variation] say? It says something about human history. I don't think it basically makes anybody unequal. Will people use it to study things that are much more sensitive, like, say IQ? I am sure people will do the studies eventually.

The point is not—there's nothing negative about doing the study. The question is what are we going to do with this kind of information, and how are we going to interpret that information?

Chakravarti's comments reflect the fieldwide belief that genomicists should be trusted to make good ethical calls in their studies on minorities. Though scientists welcome input from other experts, most are confident that genomics can handle the responsibility of explaining the most contentious variation.

Resistance to Official Standards

Despite all the strategies scientists have devised to promote minority inclusion in genomics, there is a disconnect between genomicists' adherence to the inclusionary paradigm and their support for the mandated mechanism for achieving inclusion. Since researchers seeking American funding are required by Congress to use Census categories in all publicly funded research, it would seem to follow that genomicists use the government's taxonomy for sampling and recruitment. Remarkably, none of the scientists I interviewed use the government's taxonomy for these purposes without making personal adjustments. Furthermore, only one scientist voiced support for using this taxonomy to structure an entire research program from sample design to reporting. Although the U.S. government funds all of these scientists in their domestic and foreign projects, in our conversations many scientists didn't know what the current official classifications were, and asked me to explain them. After my listing them, scientists would express discomfort with various aspects of the federal taxonomy. Perhaps this is not surprising given that the HapMap Project had already altered its use of the taxonomy with the full backing of the NIH in 2003. Yet it shows that scientists have more control over their taxonomies than the reigning policy would suggest. Each new sampling effort provides another opportunity for scientists to weigh research aims in terms of personal values, institutional policy, practical needs, and cultural norms.

Scientists are particularly offended by color-coded terms like "black" and "white." As Mike Bamshad, whose research has focused on caste variation in Bali and Andhra Pradesh,[26] expressed, "It makes no sense to me as to why I'd group a people based on skin pigmentation alone." To many, pigmentation is a superficial trait that is insufficient for representing medically relevant ancestry. Scientists also take issue with classifications that have troubled histories. Jorde said: "I don't say 'American Indian' and I say 'Asian American' . . . I get confused when somebody says you are non-Hispanic white." The term "American Indian" bothers scientists who are used to addressing indigenous Americans by the politically recognized term, "Native American." In the wake of HapMap, the term "Asian" has also come under scrutiny for glossing over location-specific variation. While Jorde was the only person to openly criticize the federal government's distinction between "non-Hispanic white" and "non-white Hispanic," others were just as confused about where the government got the idea to subdivide as such.

From the scientists who were familiar with the taxonomies, I heard many critiques of the practical efficacy of using what people in the field refer to as "Census categories." Criticizing the federal classifications on scientific grounds, Hirschhorn complained that they are too broad and might have varying relevance in different contexts:

> I was just having a call with somebody about [another researcher] who was trying to look at the association of ancestry in Latinos with a disease, and they were looking at people from Puerto Rico. But they had actually come from California, and their sort of experience of Latinos or Hispanics was all Native Americans versus European. And actually, in Puerto Rico there is a lot of African ancestry, so the markers that they used probably weren't even right.

Hirschhorn is referring to the idea that admixture levels are different throughout the Americas. Here genetic ancestry clearly trumps any social descriptor the government could provide. Hirschhorn's frustrations are shared across the field, as scientists struggle to create taxonomies that are more reflective of genetic ancestry.

Flexing his vast independence from public funding, a result of his ability to produce entirely privatized projects, Craig Venter argued the foolishness of assuming that American standards would coincide with global patterns of genomic diversity. Venter related stories of travel to distant lands in search of population isolates—societies where first-cousin marriages have predominated for centuries—to point to the rarity of finding medically interesting populations in the United States. Calling the population genomic case-control method "bullshit," he described a future when disease profiles would overtake racial taxonomy as the foremost commonsense way of understanding human similarity and diversity.

Some scientists expressed extreme aversion to the federal classifications. Spencer Wells exclaimed:

I think they are ludicrous, personally! What does "Caucasian" mean? It's everything from somebody living in Belfast to somebody living in Southern Spain to somebody in living Tunisia to somebody living in Sri Lanka. That's a *huge* range of variation. What does "black" mean or "African American"?? There's more variation in the average African village than there is in the rest of the world outside of it combined.

Even Francis Collins characterized the standards as "based in a way that is pretty hard to defend." Collins said the federal taxonomy "tends to reify the concepts that those groups are biologically different, and clearly that's not defensible." All of these positions echo the words that social critics of genomics' science of race have aired in recent years. They suggest that scientists too are critical of standardization.

Nevertheless, scientists most often criticize the categories from a scientific as opposed to social or political standpoint. Instead of citing the fluidity of racial identity or the transience of commonsense labels, scientists complain about the lack of genomic expertise behind them. Kittles surmised that Census categories are produced by outsiders to the field: "I think the Census is *whack.* I think the Census uses the federal government's categories based on a group of scientists who come together every couple of years and say these are the rules." Scientists are more troubled by the idea of outsiders enforcing classifications than they are about the possibility that *all* genomic labels, in some way, reify difference.

One social preoccupation that does register with scientists is the ineptness of American categories to properly represent the political landscape in foreign research. Discussing her research in Africa, Mountain said: "NIH is trying to gather information about whether people [are] being fair, [but] as soon as you use foreign samples or you have subdivisions within some OMB category, it does not work very well and people are less inclined to even fill it out." Mountain gestured towards a more flexible taxonomy that could meet context-specific needs. Others, too, related the importance of minding *other* governance standards, such as the official tribal classifications used in foreign contexts and local census terms. They reminded me that to get samples in foreign zones, alternate ways of representing variation must be heeded.

In most cases, these scientists are under the impression that the federal taxonomy is optional or negotiable. Instead of viewing the taxonomy as a guide for research design, they say they merely report their own findings to the requisite funding agency in the government's terms. Scientists claim to have no trouble modifying their own categories to generate tabulation in federal terms. Many give the enrollment forms a passing glance. As Shriver said, "The only time that we use those categories are to fill out a table at the end of the grant applications."

All in all, genomicists view the federal standards as unrealistic and unenforceable in lab and field. Many told me that this doesn't affect the awards process, because grant recipients can always report that they are unable to compare large populations. Rotimi said: "I think for me it is a bureaucratic process that really has no teeth. . . . People just check those things, and they come out with some reason why [they] do not have [a] group. 'It is just, I do not have control over this.' 'It was already collected samples.' How do you argue with that?" Scientists argued that recipients might also lie on federal forms. Short of checking reports against publications, no one can be sure whether federal standards are in play.

Still one cannot overemphasize the extent to which elite genomicists support the inclusionary *intent* of federal mechanisms. Despite these forceful critiques, most repeatedly declared the primacy of minority inclusion at any cost. Collins, for example, remarked:

> Here is what I think the hardest part of this is, though: If we do believe in health disparities—and they're not all on the basis of genetics; a lot of health disparities have nothing to do with DNA; they have to do with socioeconomic status, with access to healthcare, with educational opportunities, with cultural practices, especially diet, with environmental exposures which may be correlated with where you are in the world, and what particular group you are in—and we want to understand that and we want to do something about it, if you were to wipe out the current descriptors that we use for race and ethnicity in a way that fail to capture those environmental differences, those social circumstance differences, then you would essentially walk away [from addressing them]. . . .
> We don't really have an obvious alternative.

Collins's comments convey the discrepancies that the field continues to wrangle with. Like many social critics, the genomicists who are the harshest critics of government standards prove to be some of the most concerned about minority inclusion. Most scientists believe there is a way to combine study protocol with social policy. Yet their solution is for genomics, not the government, to find that way. Even policymakers, like Collins, would prefer that federal standards not drive sampling.

Self-Identification and the Reinscription of Race

Where genomicists come most directly into contact with the political landscape of identity is at the moment of subject self-identification. As mentioned, subjects are directed to mark boxes and sometimes write in responses before researchers run ancestry estimation technology on their DNA. Scientists could "eyeball," or decide for themselves, a subject's racial, ethnic, or ancestral affiliation. Instead they permit research subjects to mark their own affiliation. Scientists purposefully avoid assigning ancestry to their subjects. They told me that they prefer self-identification to registering a subject's ancestry by the researcher's impression, because it best respects the research subject's dignity—which is believed to be especially sensitive in the case of minority inclusion. As Shriver commented, "We don't want to insult the people who are being sampled, the population, by using completely academic or terms that they would not use or terms that they find offensive!" Through a policy of self-identification, genomicists see themselves as taking sides with minority group dignity and self-determination, giving voice to such groups' collective struggles for political recognition. Here is where racial practice and race as a practical taxonomy most clearly articulate. It is the social gravity of a racial taxonomy that allows research pragmatism to be guided, in the first instance, by political pragmatism and social values.

The scientists I met were entirely comfortable with this politically pragmatic tactic and spoke of self-identification purely in terms of its ethical advantage. Chakravarti described his eventual acceptance of careful public engagement around identity as a necessary part of the rigors of genomics:

> So, I think so far what we have done is called populations by whatever is socially and culturally the most acceptable way of referring to them, but that doesn't mean we are consistent. . . .
>
> I think [it] would be different if we have continual access to the populations we study. I have done some work with Old Order Mennonite in Pennsylvania; as soon as you meet the real-life human beings and you talk to them, you think of them as very, very different creatures than if you just get samples that come from a repository.

Even consistency is subsumed to acknowledging people's self-identities. All these scientists engage groups on the groups' own terms as a means of respect, though they have different levels of familiarity with and commitment to the groups they study. This policy has beneficial ramifications, since scientists are able to gain immediate access to groups and avoid creating public relations problems down the research pipeline.

Subject self-identification introduces an inconsistency into genomic sampling that can only be rationalized by the social and ethical benefits it brings. Though scientists provide subjects with a project-specific list of categories to choose from, thereby imposing a modicum of consistency on the reporting process, each lab produces its own taxonomies according to what the lab director deems socially appropriate. Instead of using one standard set of categories, as the NIH would have it, the field factors an endless array of ancestry categories. A project focused on people of African descent in the South may provide ten local subgroups, while a study comparing recent immigrants with U.S.-born minorities may provide just as many categories specific to immigrant ethnicity. With subject self-identification, scientists put their values ahead of instrumental study design in order to produce a socially beneficial way of engaging the public.

Of course, some find ways to scientifically rationalize this field's intense interest in self-identification. Elad Ziv explained how self-identification serves as a cost-effective shortcut to the kind of epidemiological information about the research subject that people like Collins desire:

> Actually, for Chinese and Japanese, [distinguishing variation] may take more than five hundred or a thousand markers, but with *one word* on a questionnaire, you can figure it out.
>
> So, self-report is not valueless even from the genetic perspective. But also, from the other perspective, self-report tells you a lot about social aspects or cultural aspects of that person, or sometimes it correlates with socioeconomic things.

Though Ziv expressed his support for adhering to U.S. Census categories in "The Importance of Race and Ethnic Background in Biomedical Research and Clinical Practice,"[27] even he prioritizes the social relevance of self-identification over standardization. Being able to account for both biological and social factors is important to a field interested in gene-environment research. Ziv comes to genomics from a background in the genetic epidemiology of cancer, yet even the most nonclinical molecular scientists are spearheading such studies.

While these genomicists devise their own taxonomies, refuse to directly apply the federal taxonomy, and calibrate their taxonomies to subjects' self-identities, they also increasingly seek answers to racial health disparities in dominant cultural terms. In fact, their sense of what constitutes "minority" and "minority health" entirely rests on the patterns of social stratification that the U.S. government aims to address with its Census taxonomy. I consistently heard scientists shuttle between their preferred continental terms, the government's terms, and the color-coded terms they avoid when designing their own studies. Despite scientists' discomfort with extramural notions of race, I found them making exceptions for the use of common lay conceptions of race as a heuristic in research that covers health disparities.

Daly explained: "We can substitute racial labels in a purely genetic study with actual hard data now, but we can't do that when we're looking completely across the board and in terms of how people are treated and what access they have had to medical care and what access they have had to early life advantages, nutrition, and so forth." Scientists like Daly, many of whom come from a background in informatics or molecular medicine, overwhelmingly supported the use of race in measures of the biological effects of racial discrimination.

From Collins I heard another variation of this racism-conscious epidemiological rationale. Collins explained that though race is not embedded in a person's biology, it is an important part of a holistic assessment of an individual's health:

> I think realistically you would want to know for every participant, okay, what do they self-identify with as far as race or ethnic group? And I think self-identification is the right answer, not what you think somebody is by looking at him across the table. And you also want to know, where did their grandparents come from? You want to know, what was their socioeconomic status and what has been their level of educational opportunity? What environment did they live in, not just the zip code but in their local environment? What are they exposed to? You want to know about their diet. You want some measure of what degree of social stress they are experiencing.

Collins and others also draw attention to race's power in the clinical encounter. They describe racial taxonomy as the language with which patients are often most comfortable.

Along these lines, Mountain supportively argued, "You could invent new labels, but you would not have any understanding of how health disparities fit in with those labels." She called the search for health disparities "a lose-lose situation," because "for practical reasons, you want to use [race]." Ravi Sachidanandam, lead investigator of The SNP Consortium and an epigenomics researcher at Mount Sinai Medical Center, argued, "This is a problem of science versus popular language, and as far as race, it is something because it has a connotation culturally that you cannot erase from your head, no matter what you do." Sachidanandam echoed the sentiment that skin color made no sense as a marker of genomic difference, but he maintained that study designs would continue to reflect target markets expressed in common lay terms. These scientists framed race as an insuperable, but manageable, linguistic challenge; race, they stressed, was a defining principle of any taxonomic heuristic. At the time, Collins and others I met were attempting to forge a large prospective cohort study that the National Human Genome Research Institute would conduct.[28] It is unclear whether those samples would have been stored in racial bins and deidentified, as in the case of the Polymorphism Discovery Resource, or openly labeled by some form of race and ethnicity.

Labeling and Reporting

As we have seen, for each genomic study that involves new samples, scientists consider how to classify their populations at the initial point of research design. During the recruitment phase, scientists may change their taxonomies. My observations of principal components analysis and ancestry estimation based on ancestry informative markers attested that population descriptors are not utilized until the data analysis stage is complete. Software programs assign numerical codes to clusters and outliers.[29] Yet after running ancestry estimation on samples, scientists must decide again how to represent their populations in their database and in data reports.[30] This gives genomicists the opportunity to rethink their initial descriptors and to take into account other issues, such as how their colleagues, scientific audience, and the wider public will feel about their taxonomies. A layered confluence of statistical classifications, geographic descriptors, and population taxonomies is apparent in journal articles, which often give finer-resolution ethnic or self-reported categories in the methods section while using broader racial terms in the body of the article.[31]

All of the scientists I spoke with said that for data reporting, they attempt to square their study classifications with descriptors that are socially, if not politically, relevant. Speaking generally, Stephen O'Brien said that "the labels for the major ethnicities are rather straightforward, and they change depending upon the sensitivities of the communities." Reich reiterated the point by sympathetically emphasizing the political transience of labels: "There has been a history of changing the word that African Americans call themselves over time—'Negro' and 'black' and 'African American' and 'colored' and so on. . . . People have been constantly changing the word because every word gets loaded, and then they move to another word over time." In fact, the extent to which scientists reflected on the histories of the labels they use surprised me. After showing me his favorite race-troubling slides—images of two highly overlapping Gaussian curves indicating the biogenetic similarity between African and European Americans—Jorde confessed that he avoids terms that "incite negative feelings." He and others recounted the problematic history of terms like "Caucasian" and "gypsy," taking issue with social labels derived in the efforts to distinguish some people as superior and others as inferior.

Mountain expressed the common belief that genomicists should adapt to their subjects' political context and try to think in their shoes:

> What I do is I imagine that somebody with that, who might get that label either for themselves or someone else, is reading it and can understand it. Are they comfortable with it?
>
> So, when I studied different populations in Africa, I'd think, "Well, what if they would have read this? Will they be comfortable with it?" And that is hard. Work is hard enough anyway!

Since some scientists feel that noting the sample population's self-reported identity in the methods section of articles is a good way to "give voice" to their subjects, they avoid using purely genetic terms. On reconciling reports with subjects' viewpoints, Wells insisted:

> We don't tend to relabel people, because that is not what we are in the business of doing. But you know, in some cases, you will sample people who report to be unadmixed and turn out [to] be quite mixed with surrounding groups. And that will force us to reconsider their population history.
>
> But we are *not* going to change somebody's self-identified ethnicity.

These scientists attest instead to layering self-reports and genetic reports throughout communications. Rotimi illustrated:

> At the beginning stages, I try to make sure that everybody knows where these people are really coming from. And then later on, I can say West Africans.

But it comes to a point where you now want to look at the people who have recent African origin, because in the end I always tell them we are all Africans beneath our skin. But at some point, when we want to describe a diaspora, for example, which is the recent migrational patterns, then the concept of "Africans" and the "African diaspora"—sometimes the word "black" and just given the way we have defined that—captures that.

The extent to which genomicists willfully press their data into socially relevant categories so that their research can be politically acceptable was most evinced when I asked project directors to explain what constitutes a "best" descriptor. Contrary to a strictly data-driven response, all whom I spoke with thought through the question with political considerations at the forefront of their minds. Most explained that geographic terms like "sub-Saharan African" and "East Asian" make for superior labels because they are specific yet flexible, good at imparting the primacy of the environment in genomic processes, and less politically charged than other kinds of terms. Scientists explained that geographic terms often coincide with self-identified cultural monikers, but in the instances where self-identified terms are publicly contested, geographic terms can provide a neutralized alternative. For example, scientists prefer the terms "African American" and "European American" to "black" and "Caucasian." Scientists also rely on the latest versions of ethnolinguistic terms when geographic terms are insufficient. For example, the HapMap Project distinguishes between Ibo and Yoruba, because they are communities that cohabitate in Ibadan, Nigeria, yet form separate social groups. Likewise, scientists working on migrations in a narrow geographic territory defer to community leaders for preferred labels.

Although the research directors I spoke with have the ultimate say in how their populations will be communicated about, such autonomy does not mean that scientists invent taxonomies in a vacuum. Genomicists confer about commonplace classifications and share publicly effective taxonomies. Through collaborations, informal dialogue, and peer review, scientists communicate preferences and advice. In doing so, they create a patterned discourse and convergence of values about racial communication, which can be referred to in disputes with critics.

Hirschhorn, for example, told me about his collaboration with Goldstein on a 2008 meta-analysis of European population studies and the deliberations they produced[32]: "We had discussions of [whether] this was the work we actually wanted to be doing—is this work we want to be *publishing*? And we actually talked about these issues, and how do we *want* to talk about—we actually spent a fair bit of time when we were describing different European groups [and so on].” Reflecting on his own terms and his interest in shaping the discourse through collaborations, Shriver similarly said:

> I may have used the word “African” in some earlier papers—people evolve in how they see the concepts themselves. But I never say “African” now. When we are talking about the ancestors of African Americans or Brazilians or black British, “West African” is a much better way to describe them because those are the areas from which the enslaved persons were taken in the sixteenth through the nineteenth century.
>
> Well, I just finished editing a paper with [an admixture collaborator], and I changed all the “Africans” to “West African” (I did not change the “Native Americans” to “Indigenous American,” but I usually would—I think there are different ways to see this, but “Indigenous American” is preferred in this context because “Native American” is also a political group).

Shriver continued by giving another example, this time an ongoing dialogue with a longtime research partner: “Also, you don’t want to allow for inappropriate generalizations of your results, which is what happens when you say ‘African’ or when [the pharmacogenomics collaborator] says ‘sub-Saharan African.’ I have talked with him about this many, many, many times!” Like others, Shriver was certain that though he did not have direct control over research he did not conduct, best practices would emerge through collaboration. These best practices would then influence individual research habits.

The scientists I interviewed also spoke about coming to consciousness through journal refereeing and having their own work refereed. Scientists’ stories of peer review centered on how this structured dialogue narrows label options and increases researchers’ awareness of broader community interests. Mike Smith described a give-and-take exchange with other genomic elites that set off a chain of refusals to use the term “Caucasian”:

> A colleague over at Hopkins . . . had used the term “European” . . . and then elsewhere in the paper she had used the term “Caucasian.” And, I wrote to her as I reviewed the paper, and I said, “Look, you gotta pick one here”. . . . I said, “My preference, and what I have settled on, is ‘European American’ and ‘African American.’ It’s *descriptive*.”
>
> Years ago, Ken Kidd yelled at me for using the term “Caucasian,” and he said, “Those are from the Caucasus! That’s over in Russia!” And, I just said, “Fine, I’ll stop using that.” And by and large, that’s worked.

The collegial dialogue described here keeps the field integrated and attuned to socially accepted taxonomies, creating cohesion among people differently positioned in society. It shows that writing and reporting form a critical moment when scientists at different labs connect with the social context of what are considered appropriate labels, thus streamlining the field's taxonomies and prepping the field for public interface.

Yet scientists also describe feeling constrained by collaborative work in ways that make them uncomfortable. Reich gave the example of a cohort study with which he had written a couple of papers: "It's a *huge* collaboration with all these internal rules and paper-writing rules. They call African Americans 'black' there. In those papers, I had these fights associated with that. I've tried to call African Americans 'African Americans' the whole time, and they want to call them 'black' because that's their rulebook." Many scientists who work in the Northeast complained that projects that sampled in the South and Midwest were more permissive of color-coded terms. Reich ruminated: "These people just talk about things that everybody here would be uncomfortable with. [Maybe] it is more a reality there. Or maybe it's more behind the time. . . . Maybe it's more ahead?" Reich is upset by the insensitivity of his biomedical collaborators in using what his peer group considers to be outdated terms; yet he is also puzzled by the continuing favor these terms find in communities where minorities concentrate. His comments show the normative quandaries that arise as scientists attempt to adjust taxonomies across geographic, cultural, and institutional divides.

Against Standardization

As alluded to in Chapter 3, biomedical and science journals have set policies that try to get scientists to define their terms and to use specific taxonomies.[33] None of these policies have been followed up, implemented, or enforced.[34] Moreover, genomic research is continually testing the bounds of race, ethnicity, and prior population taxonomies. A definitive human taxonomy is always "just around the corner." As a result, genomics operates without a labeling or reporting policy.

The scientists I spoke with expressed great confidence in the social and scientific value of the labels they devise themselves. Such confidence might lead to a variety of labeling standards. On the one hand, scientists could view standardization defined by genomic terms as a benefit to the public. They could unite to create a field-sponsored taxonomy to be implemented by the NIH, the International Conference on Harmonization of Pharmaceuticals, the World Health Organization, and other health organizations. They could also infiltrate panels determining labeling standards so that they could steer larger policy efforts, such as the redistribution of federal resources and racial reparations. On the other hand, disagreements with each other's standards, government standards, and lay taxonomies could disenchant scientists with the labeling process. This could result in a stronger interest in stripping data of social signifiers—even geographic categories—or a final move to symbols, as in the case of haplogroups. It could also lead to scientists bowing out of the debate over race and further turning to an "every lab for itself" protocol.

Each of these permutations would bring a unique balance of benefits to genomics—some fostering researcher autonomy and others stabilizing the field's guidance of public thought. In fact, these lab directors and project leaders proved to be remarkably unified: just as they valued their autonomy in sampling, they were vehemently against labels standardization. All but one scientist I spoke with said that researchers should create sample taxonomies based on each study's unique research questions, obtain as many layers of information about subjects as possible so that many labels can be circulated about the same data, and report with classifications that are beneficial to the groups under study.

Interestingly, it was the most race-averse scientists, who would be expected to support the standardization of categories that clearly contrast racial taxonomy, who expressed the strongest antistandardization views. Rotimi, for example, said that a detailed description would be superior to standard categories: "Just describe what you think you have, and what you will use it for in your study." Kittles reiterated that in time genomics would evince the speciousness of the labels used in studies:

> I don't think that we have to get together and have a meeting about the right label. In the end, I think if you allow geneticist to do their research, they will show through their research that [the racial label] is not clear. If you stop them, they will not. I don't even think that in this country there's a way of stopping people from using those labels. I guess you could prevent further gathering of data using the racial categories, but people will continue to use them as long as they can.

He brought it back to the issue of self-identification:

> I think that there shouldn't be certain rules because there is no one set of labels. It depends on the research and the design of the study and the project and the

questions that are being asked. So one of the things is we should be very clear. We should just ask people who they are and how they identify themselves. That is the first thing: self-identify.

Mountain resounded Rotimi's charge to define terms:

> I think, better than just standardization, we have to explain exactly what they mean by any label. Do they mean people self-identified? Do they give people an opportunity to give multiple labels or not? Even do they give check boxes or do they have people fill in blanks? You know those—there is not any absolutely ideal world. Whatever label you use, you say how you came up about using that label.

Mountain found it frustrating to read papers where study authors use continental terms without further explanation. She cited her then recent article with bioethicists Mildred Cho and Pamela Sankar, which encouraged genomicists to define terms with as much detail as possible.[35]

In discussing standardization, interviewed scientists again derided oversight. Some ironically chided the "commissions" and "committees" that, as O'Brien put it, "just sit around and make all these pronouncements." More than anything, the scientists I spoke with were afraid that standardization would elide the self-identified terms subjects may hold dear. Because all leading scientists work with vulnerable populations, scientists sacrifice analytical acuity for subject self-determination—an ethic I have discussed earlier.[36]

Only one scientist, Esteban Burchard, made supportive comments about fieldwide standardization. Using an objectivity rationale that one might expect of scientists were they not so concerned with the ethics of public engagement, he said that standards were sound science: "It can't be the Wild West. I mean, this is just good science. It just makes good science sense. . . . You can imagine trying to buy something, and you say, 'I want a kilo of this.' If everyone is using a different measure of a kilo, it gets crazy!" As he has done in the literature, Burchard privately supports racial standardization as part of a minority inclusion program. Burchard most clearly believes in the strategic essentialist notion of racial heuristics until justice has been served.

Most typically, however, these scientists view genomic taxonomy as a fluid process. As Jorde articulated, "I think we are always trying to sharpen our terminology, refine it, and make it as accurate and descriptive as possible." He said:

> Now in the forensics world [that I work in], police and so forth, they use ["Caucasian"] all the time. And the *New England Journal of Medicine* prefers "white"

and "black." So, they use the term "whites," and some people would have trouble with that. So, there is *no* descriptor that works perfectly all the time for *anyone.*

Jorde and others acknowledge the "Wild West" climate of the field but assume that it is necessary if scientists are to maintain their pragmatic autonomy. They are aware that the field benefits from keeping a wide latitude for negotiating politically salient labels. In covering a range of racial and ethnic bases, scientists are able to use racial taxonomy to communicate with the public, try on various labels and avoid reducing their work to one taxonomic type, but also to satisfy many political standpoints at the same time. It has not been documented how various labs integrate layered accounts of race, ethnicity, grandparent ancestry, and the like. Nor is it clear what weight each variable is given in individual studies.[37] However, it is plain that autonomy permits genomicists to deal flexibly with the ethics of their science as it unfolds.

A Taxonomic Practice to Amend a Field

Genomic research is based on a pragmatic principle that incorporates social taxonomies as a method of including underserved groups. In sampling, scientists alternately revile and defend the necessity of using race. They collectively seek alternative classification systems so that they can move beyond racist or Eurocentric science, yet their continentally stratified alternatives reflect a priori categorical frames that the public understands and that are meaningful in lay identity constructions. Research directors are vehemently opposed to the standardization of categories across the field, to the point where they resist government oversight. In applying labels, they are generally more interested in personal consistency than collective consistency; however, notions of what is politically appropriate inform the entire research process. While disapproving of federal standards, many welcome subject self-identification, even if that invites lay taxonomies that unabashedly refer to race.

Still scientists are also ambivalent about specific racial taxonomies and conflicted about the depth to which lay categories should penetrate their research. Clearly, their most favored taxonomies are in constant transformation with respect to their immediate social contexts. Genomicists value cultural sensitivity, minority inclusion, and subject self-identification, but they stake out an autonomous space to deliberate and direct the politics of their science.

Scientists are highly concerned about the social consequences of their choices, because as the publication record shows, the groups they decide to sample become the reference populations by which future research and applications are measured and compared.[38] This is as true for sample populations collected by an individual lab as it is for sampling in large-scale projects. Researchers who for example set out to find forensic markers specific to Africans or Asians, or who frame health as different for blacks and whites, end up reporting variation in those terms. Other research teams refer to those studies and use them as a springboard for new investigations. When authoritative labs and projects are the source of such findings, the protocols and resulting taxonomies have even greater ripple effects. One sequencing project's sampling taxonomy can condition everything from pharmaceuticals and diagnostics to forensic and genealogy kits. Through these applications, genomics markets biology in ways that encourage the public to view themselves through a specific lens of difference. Furthermore, well-connected scientists share and legitimize taxonomic strategies, creating a limited field of practice for the rest of genomics. Thus, taxonomies fashioned for elite labs eventually affect conduct in other labs, scientific contexts, and the broader society.

By virtue of the simultaneously external and internal, concrete and cognitive, nature of knowledge, "acts of *cognition* are, inevitably, acts of *recognition*, submission" to a preexisting scheme.[39] In other words, "practical taxonomies," as Pierre Bourdieu calls human systems of difference, are at once objectively induced and subjectively produced, because we think with, apply, and reinforce them in a variety of contexts. Racial categories become obdurate features of scientists' everyday lexicon, because they are "already socially salient."[40] There is a long and visible history of people organizing on the basis of race; certain groups are recognized by the state; and many advocates in health, medicine, and beyond consistently succeed in "linking justice arguments to biological difference claims."[41] Drawing on Michelle Lamont and Virag Molnar's phenomenological explanation for group classification, Steven Epstein also suggests that a performative dimension fosters race's pertinacity. Scientists rely on race because they *enact* race in their everyday lives. Racial categories span scientific, bureaucratic, and personal aspects of their everyday experience.

Racial taxonomies are characteristic, practical taxonomies, because they draw on conceptual binaries that are perceptible in biology and society and entwine with long-standing, reified *social* hierarchies. The drive to use racial taxonomy permeates scientific practice even as scientists attempt to alter that practice. This analysis has explored how identification and classification interlink as scientists manage biological utility, public clarity, social relevance, and even cost to produce the staple databases that populate the field. These patterns in taxonomic practice reveal common strategies used by the field's key decision makers. It is also useful to examine race-positive habits that are mobilized by the field collectively. This would include the field's trajectory of engagement with the race concept, as well as scientists' shared vision for genomics. How does the field seek antiracist racialist alternatives? If we are attentive to the objections that people raise to knee-jerk deployments of official standards, we may learn more about the way science activism is being cultivated in this privileged arena.

6   Activism and Expertise

LOOKING AT GENOMICS TODAY—a field in which examinations of race are commonplace, and calls to use technologies to decipher race are routine—one might guess that genomics was from its inception destined to be the new science of race. Yet the race-positive science that we see today is the product of a very specific configuration of expertise. Examining the ways in which elite genomicists amplify their role as public reformers, I discuss how scientists' shared belief in their capacity as public stewards leads to their confidence in establishing genomic taxonomies as a better alternative to race.

As we have seen in earlier chapters, genomic race-positive science is in part conditioned by partnerships that scientists forge with social scientists and humanists. Not only do genomicists sit on a number of committees and panels dedicated to collaboratively managing racial reform, but they also coauthor op-eds and proclamations from an interdisciplinary perspective. While their methods may be collaborative and may appropriate the terms of social research, these scientists still hold onto their claim to biological objectivity. As such, genomics becomes consolidated as the apogee of expertise in a wider field of expertise on race. Thus, I also explore how interdisciplinarity between natural science and social science buttresses the field's social and symbolic capital while simultaneously securing its position as the ultimate authority on scientific matters of race.

Genomicists seem to signal a departure from science-as-usual, particularly the characterization of scientists as cold, distant, arrogant, and impervious to the social world—a depiction that has dominated humanities' and cultural studies' critiques of claims to "scientific objectivity" during the so-called science wars.[1] Contrary to the image of the scientist within the fortress of the "citadel,"[2] genomicists work to publicize their values and to garner attention for stances that they share with nonscientists. Scientists no longer hide behind a veil of impenetrable objectivity but rather wear their politics on their sleeves. Genomics functions in some ways like a social movement or an advocacy group. In their attempts to come to society's rescue, members of the field bring what Steven Epstein has called the "tacit coalition" of interdisciplinary, interinstitutional expertise into full visibility.[3] Visibility, consensus-making, and debate all entwine to further feed the field's status as ushering forth a panacea to the world's racial problems.

We begin with scientists' own articulations of their motivations and goals, because these speak to the field's range of engagement with the broader society. As others have shown, in this hybridized arena where elements of the Triple Helix—and increasingly basic research and translational healthcare—are indistinguishable, scientists and the values they hold dear maintain great influence.[4] Hailed into dizzying public debates, with increasing knowledge of the stakes of their research, scientific elites develop norms for self-fashioning that propel the field's broader reputation as a leading science. Their investment in proactive, *activist* science also shapes the velocity and momentum of scientific development.[5] At a time when reputations of personal honor and trust drive the field's ability to deliver on promises and create new billion-dollar ventures, motivations matter; *ethos counts.*[6] An activist ethos facilitates an interdisciplinary expertise that in turn greases the wheels of genomics' ascendancy. I conclude by discussing some of the limits of this science-based activism.

The Changing Role of a Scientist

Genomic elites fashion their expertise in multiple ways. One of these is through charismatic publicity. Steven Shapin's recent work has illuminated how charisma drives modern science and expertise through the "moral warrants" that the public requires of scientific practice.[7] Indeed, the desire to actively shape society's moral frameworks has pervaded the upper echelons of genomics since the earliest ELSI deliberations of the Human Genome Project. These scientists collectively fashion themselves as ethical avatars—pioneers of the good life, leading society toward a better future.

Reflecting on the scientist's role in society, Craig Venter summed up:

> The ideal is that we all try to use our power, our positions, if we have any, to try and influence the world around us. For me it is very hard to be in truly a modern society looking at the past fore-history of our species and not to be upset, be ashamed of it, extremely bothered by it.

Francis Collins was similarly heartened that genomicists' efforts to get the public educated about "the meaning of the genetics and race interface" are working and that the majority of the scientific community is now invested in getting race "right." Since the publication of the draft map of the human genome, both scientists have consistently appeared on television and across the news media to present their views on race. Their stance is a far cry from the "moral equivalence" that scientists have assumed in the past.[8] Genomicists express assurance that their values about race and equality can serve as an example for others. In essence, they fashion themselves as moral models—members of society who demonstrate how to live mindfully about *meaningful* differences.

The prevailing sentiment in the field is that scientists should go public with their knowledge and generate as much social change as possible. Many think that genomics should increase its visibility and be at the forefront of the public debate over race so that people will learn to think about difference as genomicists do. More generally, they insist that genomic knowledge is a special kind of social expertise—a true knowledge of human variation—and that the genomicist is a special kind of ethical actor, the Prometheus of that truth. Even the three people interviewed who complained that they did not have the proper training to deal with social issues called for better sociological pedagogy in the field. These scientists do not regard retreating from social issues as an option. First and foremost, genomicists characterize public education and community intervention as a responsibility of a science that overwhelmingly depends on public support. Indeed, promises for public engagement are often written into genomic grants, and are very seriously weighed by review boards. As Stephen O'Brien reasoned:

> [I]t's the public who supports our work. It's the public who benefits from it. It's the public who evaluates it and needs to understand it. We tend to think, some scientists tend to think, "Well, if our peers like it, it must be okay." Well, I think we have a higher obligation than just getting the five or six scientists who happened to be interested in the minutiae that you happen to be studying at the

moment. You need to understand who's paying, who's writing the checks—the taxpayers, the community, and the beneficiaries of the research that we do.

He enjoined others in the field to publish and give talks in more mainstream venues, presenting public science as the ideal end of genomics. Terminology like "a higher obligation" and "duty" was rampant in scientists' free thought on their roles and responsibilities as public advocates and educators.

In fact, some genomicists feel that scientists who are uncomfortable with going public shouldn't be involved in the field. Rick Kittles remarked:

> I think that if you want to be a successful scientist and you want to contribute, you have to be very conscious of what you're saying and how you say it. And you should be able to articulate what you are doing to the lay community in a fashion that they understand and appreciate, because if you can't do that then you should probably stay on the sideline.

Scientists like Kittles stress the power that genetic metaphors hold over the public imagination. They believe that *they* are responsible for changing those metaphors. Several people referred to their job as "framing" the science. Because genomicists think that knowledge about the genome is going to play an ever greater role in society—many pointing to the eventual restructuring of society around genomic profiles and risk assessment, as Paul Rabinow argued when he forecast the rise of biosociality—these scientists see their interpretive role in public as growing by leaps and bounds.

Aravinda Chakravarti mused: "I know in this society that, and many societies, scientists are viewed as this machine that are completely above and oblivious, and that's just not the case at all. The nerd-of-the-scientist has gone." In contrast to the "nerd" with no sensitivity to human concerns, he sees the scientist today as an activist, an equal partner in the policymaking process. With respect to the scientist's responsibility for racial knowledge he told me, "One thing we have to do is get the appropriate laws on the books."[9] This proactive political engagement has been built into the field in its simultaneous move from colorblind to race-positive science and a high-profile public health approach.

Scientists credit their human variation studies, and opportunities to reflect on the racial implications of those studies, with fueling their sense of responsibility for building activism from expertise. Esteban Burchard said that through his public engagement, including numerous visits to community centers and a 2010 spot on the *Tavis Smiley Show*, he has grown certain that it is up to scientists themselves to "make sure politically, and as a global community, that we have safeguards in place that don't allow for the manipulation of this information." Esteban Parra illustrated how he monitors mainstream reporting on admixture to make sure his version gets airtime, so that he can "directly tackle the issue of oversimplification and misrepresentation." His and many others' stories point to the way the development of expertise about race can shape the overall outlook of a field. First, members of this field share in a proactive ethos of public responsibility, where it makes sense to consider inclusion and health disparities at the outset of research, and weigh scientific intents with respect to social implications. Second, they share a confidence that their research is good and beneficial, and that gaining public support is simply a matter of proper communication.

These perspectives on what it means to be a scientist intersect with the personalization of racial inquiry. As Joanna Mountain illustrated: "For me it is a huge opportunity to think very pragmatically about a lot of these issues, and it is exciting, but also I have this role of—I want to make sure the science underpinning the product is really of high quality." Mountain sees herself as a bridge between several academic, industrial, and public communities—namely, 23andMe, Stanford University, the general public in the Silicon Valley, and the East African groups with which she has worked. She and the others emphasized their primary commitments to their kin, ethnic communities, and broader social networks outside the lab. They described their work as a dialectical process where they consider the moral and ethical values they have accumulated over their lifetime in conjunction with the immediate issues presented by scientific research. Again, a subjective approach is not shameful but rather lauded.

Though these scientists are optimistic about their ability to challenge notions of race they find problematic, many add that they would always follow their values over a science running in a direction they see as unfit. Thus, they feel they merit absolute trust from the public. Recall Chakravarti's critical observation on the roles he holds in society: "I cannot believe in things that my science says is not true, but the reverse is also the case. Just because I could do a piece of science, if I think it's socially—really, personally—not acceptable to me, I would not do it." Many whom I spoke with insisted that every scientific decision is, and should be, an ethical one. Whether engaged in subject recruitment, population labeling, or public engagement, scientists work with a belief that in the ultimate instance, values trump rational science. Again this doesn't mean that they feel their finished product will be skewed, but rather that as genomicists they hold the power to make the right call in each situation.

Stewardship was a common theme in all my conversations with genomicists. Scientists spoke of a dual stewardship—stewardship of biomedicine and stewardship of the public. Generally, they gave the impression that it was their responsibility to educate clinicians and those less versed in genetics so that they would avoid, in Venter's words, "applying mistreatment to just as many people as they think they might be helping." As Sarah Tishkoff explained: "We typically would refer to ancestry. But I would think, as I said, that the more clinically oriented people probably are not aware of that discussion as much. And maybe they just don't have as much of an understanding of the population genetics and the history and patterns of diversity and things like that." On stewarding the public, David Altshuler gave the example of working on the HapMap population guidelines. He described the intricate dialogues, which slowed the project yet put scientists in the right ethical standing so that they could bring forth a new paradigm of population representation. Despite getting "scooped" by HapMap collaborator Perlegen in publication, he and the other HapMap scientists said it was well worth the struggle. Altshuler and his colleagues want to see these policies reverberate through the field and trickle into the public discourse; therefore, they have moved forward on high-profile projects like the genetic genealogy television program *Faces of America*.

Minority scientists also describe a kind of experience-based stewardship of the field itself. Burchard insisted that he's been thinking about the ethical issues of race since he was born and is thus well positioned to bring the field and public closer to race-consciousness and full accountability for racial inequality. Kittles ran through the list of stereotypes that he thinks other scientists operate with, to set himself apart: "All blacks do not like chicken and watermelon! All blacks do not have all the same Duffy allele!" Minority scientists model a more activist public engagement in the hopes that their contributions will convince the field that fixing race will enhance its capacity to effect social change.

The stewardship genomicists present bespeaks an elite position for genomics among other forms of expertise. Thomas Gieryn describes such positioning in terms of boundary work: the construction of a figurative boundary between science and nonscience in order to monopolize expertise, protect a science's autonomy, or expand a science's role in society.[10] According to Gieryn, scientists draw a marker around their expertise by reserving characteristics of objectivity and rationality for science alone. For boundary work in genomics, elites attribute the most comprehensive rational understanding of human variation to genomics specifically, as opposed to science writ large. At the same time, genomicists include subjective qualities in their epistemic toolkit. Though on the surface it seems that the boundary between genomics and other forms of scientific expertise is blurred, the combination of authority for scientific objectivity and social subjectivity posits genomics as a uniquely bulletproof entity, with its superior combination of methods and its reputation as the human science par excellence.

Taxonomies of Truth

From my conversations with scientists, it seems that a belief in one's own moral merit and social accountability leads to credence in the field's humanistic benefit for society. Speaking about global genome projects, Georgia Dunston confided to me: "It is a very exciting time in sorting out the error from the *truth*, so we can take the facts that we can build on to achieve the outcomes that we desire." Like others, Dunston interprets genomic work as one of *truth discovery for social engineering*. She sees moral decisions, scientific research, and policymaking as intimately connected, and bases her policymaking role on this belief.

Some express their relationship to genomics even in terms of a calling. Mark Shriver said: "I had this epiphany! I don't think it was God who talked to me, but something else. . . . I started to think: how do I follow this [calling] further and find out where we are going to go?" All of these visions suggest that scientists are confident about their research, because they link it to the bigger-picture issues they see facing humanity. Hunting for knowledge about humanity is envisioned as a search for self-knowledge and social knowledge.

Likewise, no matter how they define race, scientists are confident about the global impact that genomic taxonomies will have on humanity's concepts of relatedness. Venter remarked:

> So, within the decade, I predict that there will be a different classification system.
>
> Our goal over the next decade is to do ten thousand human genomes. Out of that we will evolve a new classification system. Some of these will have practical benefit because people will—in the future, as people grow up with their genetic code, if you have it from birth there will be social identifiers that sort of peep out. They will become part of the common culture, and those will be the new identifiers. Maybe they will be dating categories because you know that you are a lethal carrier for this. And so you wouldn't want to date anybody else, or you certainly wouldn't have a child with anybody else.

There are going to be categories, that we can imagine now, that will become the major social constructs.

Hopefully, that will take us away from this ridiculous system of classification by skin color and physical characteristics.

Mike Smith, who believes that race has a genetic basis and will only be more clearly proved by new technologies, agreed:

I think that the field will naturally come to some standardization over time. I think that some of the public is going to hate this, but I think some degree of those standardizations are going to come out of genetic analysis now. Because we are typing half a million or a million SNPs on large numbers of people, and the structure of human populations is rapidly becoming very apparent.

These scientists believe that some form of DNA-based biosociality will soon be the fundamental social order and that genomics must actively shepherd society toward that future. In fact, their vision goes far beyond Rabinow's characterization of biosociality in terms of "disease risk" sociality. Instead genomicists see the most basic units of code as the future organizing principle of difference. Individual SNPs and brief sequences of noncoding material that contribute not to life-threatening but rather to mundane life processes are the pivots on which future social interaction will turn.[11]

The new science of race is built on the certainty that genetic ancestry will evince a more robust taxonomy than prior morphological systems. As Venter put it:

I want a classification system that people can know: "Here is the 25 percent purine that's based on your genetic *code*. And you have something that you should be [paying] attention to in terms of *risk*."

And guess what? When we look at those bins, I will be the most surprised person on the planet if any have been a homogenous group of any kind.

White males get prostate cancer. Black males get prostate cancer. Maybe black males get it at a slightly higher rate. But it's not exclusive to black males.

Pui Yan Kwok sees pharmacogenomics as leading the way:

I think [pharmacogenomics] is quite the place where race and ethnicity will fade pretty quickly. . . .

Right now there are about eight or ten genes that we scan for all the Jewish kids because they could be carriers of a bunch of pretty depreciating diseases.

And I bet in the future, when we can do so much more for the same price, then [race] becomes less and less important.

Kwok criticized how tests are prioritized for different populations instead of given to all members of society. Scientists like Venter and Kwok described a move back to a one-size-fits-all healthcare system as their ideal; by that they mean a system that is uniform in the sense of equal diagnostics and equal treatment for all.

Even though scientists believe in allowing subjects to self-identify their race or ethnicity, they envision a day when genomic taxonomy will make that practice obsolete. Joel Hirschhorn explained how scientists currently have to remove "genetic outliers," or people whose self-identity contradicts their genetic ancestry. He said, "the more you can learn about why that difference is there, the more you can discard the label essentially"—although he recognized that labels would hold fast because of political histories.

Many scientists speak about their job as guiding a new phenomenology of race. Famed evolutionary biologist Alan Templeton called self-identity and perception of others' identities "the first strata" to be penetrated by genomic technologies. David Hinds mused:

> People have this sort of built-in programming that wants to sort of put things into categories. And that's how it seems to be a way that people want to organize the world. There are probably good reasons why, in many of our everyday things, it's a very useful thing to have that built-in circuitry for classifying things. It has its good aspects and its not-so-good aspects.

Spencer Wells affirmed this point on a broader cultural scale: "It is an adaptation, xenophobia: you can tell who you are related to and therefore who you can trust." Shriver echoed this naturalized view of labeling taxonomy with a neurological rationale: "Humans are very fixated on faces. We have a special part in our brain, in the amygdala, that helps recognize faces and helps us interpret emotions and things." At the same time, he said that knowledge about the phenotype will teach us not to fear it:

> There is really nothing to be afraid of with ancestry. I think people have gotten to that point that the government is not so afraid. Maybe they see there is no hope in preventing people from knowing that genetic ancestry is a part of what we call race. Maybe they have accepted that with skin color now, I guess. Facial features are still off into the future, but they are all part of just what makes

people different from each other. To me, it is not going to reify anything that it might help illustrate. In fact, it is a way to demystify those differences that we see among people.

More than a new science of race, genomics in this vision is posited as that which can lead society beyond "built-in programming" and social "fixations" to a *new humanism*.[12] The suggestion is that armed with genomic knowledge, people will learn to perceive the differences that really matter to life itself.

Setting the Human Record Straight

The publication record bears out the interest genomicists have articulated in producing a definitive genomic taxonomy to replace social classifications. Testing race has become a form of science activism that allows genomicists to use what they know best to offer a foundation for social change. As shown in Chapter 3, at the start of the millennium genomics began releasing a volley of studies designed to measure race specifically. We saw the first of these in a study by David Goldstein and researchers at University College-London and Oxford claiming that common race categories did not match clusters produced around drug response genes. These statements made waves in the mainstream media, in part because the scientists themselves were growing famous. Across the pages of major newspapers, the translation of Luca Cavalli-Sforza's *Genes, People, and Languages,* in 2000 was also garnering attention for its message that the genome carried "the lesson about 'racial' differences—which is that, in the main, there aren't any."[13] Though the 1999 rerelease of his 1971 textbook *The Genetics of Human Populations* still had a chapter on "Racial Differentiation," the 2000 trade book opened with a chapter on the *illegitimacy* of the race concept. Cavalli-Sforza argued that race was a viable optic for animal variation but that it was unhelpful to the scientist studying human variation. He reasoned that there were indeed human subspecies populations, groups of individuals who typically bred only with each other, yet these numbered not four but thousands. Accordingly, avoiding the concept, and replacing it with genomic taxonomy, would be a no-brainer for the field and an indisputable fact for the world.

What has followed has been a series of tests run on specific collections like the Jorde lab's collection and the Human Genome Diversity Project samples. At the University of Utah, Lynn Jorde, Mike Bamshad, and colleagues conducted a study in 2003 to see how genomicists could sidestep using "a proxy such as place-of-origin or ethnic affiliation . . . [and] use of surrogates, such as skin color."[14] Similar to the University College/Oxford study, they created a panel of genomic markers that could allow a computer program to assign a genetic continental ancestry profile to a sample, with no previous knowledge about the individual. With one hundred *Alu* markers, these scientists found that they could confidently assign over five hundred individuals to "Africa, Asia, or Europe."[15] Still the team posed two caveats. First, the groups that were assigned an African origin always formed a northern and a sub-Saharan cluster. Second, to produce clean continental assignments, they had to purposely sample from well-defined populations. For example, inserting South Asian samples into the sample pool undermined the clustering.[16] This novel alternative to surmising the race of the research subject betrayed the field's attachment to a continental framework while at the same time positioning the field as seeking an alternative to this most basic shared assumption.

Bamshad took these findings to the public in *Scientific American*'s June 2003 cover story, "Does Race Exist?" The article's headline continued, "If races are defined as genetically discrete groups, no. But researchers can use some genetic information to group individuals into clusters with medical relevance." Written with science writer Steve Olson, author of the pop science book on genetics and race *Mapping Human History*,[17] the article appealed to the man-on-the-street's common sense about race even as it delivered the counterintuitive message that there were no easy answers. While acknowledging that a patient's race could give reliable evidence as to whether the individual was likely to have a disease like sickle cell anemia or cystic fibrosis, Bamshad and Olson argued that not all race-related genes had significant health effects. They added that superficial similarities did not always correspond to genetic similarities, as in the case of some sub-Saharan Africans and Australian Aborigines who share the same skin tone but are otherwise evolutionarily distant. Yet a color-coded map accompanying the article illustrated the global distribution of *Alus* in terms of a tripartite gloss of bronze, gold, and green corresponding to the continental divisions between Europe, Africa, and Asia. Only a strip of Northern Africa, colored gold to match Europe, matched genetic profiles with a population outside of its continental borders. Thus, the article left intact the notion that the evolutionary record proved continental distinctions important and that race could be used as a proxy for medically relevant ancestry, even as it attempted to upset commonly held assumptions about race. In an interview with the Associated Press, Jorde called this a situation where "the happy message" of indistinct boundaries had to be tempered by the importance of race-ancestry correlations.[18]

Bamshad has since teamed up with Utah colleague Stephen Wooding, and Genaissance's J. Claiborne Stephens and Benjamin Salisbury, to review where the field is with its ability to discern lay race from DNA samples. These scientists have provided examples of several population studies that have indeed determined the continental origins and likely racial identity of samples. On the one hand, they have shown that such experiments only work with markers judiciously chosen for "ancestry informativeness," like those predominant in one continental population, and only when applied to samples taken from well-defined populations. They also have warned that markers responsible for visible differences like skin color are most likely subject to natural selection in a variety of populations and are thus a red herring of distinction in distant ancestral terms. On the other hand, in an analysis of 63,724 variants these scientists have found that most are common in only one continental population. Though they advise researchers to gather as much ancestral detail as possible, they say that each study will require considerations of race anew.[19]

Meanwhile, in response to the 2002 *Science* study on Diversity Project samples, Svante Pääbo, the director of the Neandertal Genome Project and the Genetics Department at the Max Planck Institute, with his student David Serre, has shown that sampling from well-defined, or geographically distant, populations produces distinct continental clusters, whereas sampling from populations distributed evenly across the globe reveals clinal patterns in variation. With richer coverage, Serre and Pääbo were unable to assign continental ancestry to individual samples. This comparison makes explicit just how much a study's outcome depends on the way scientists choose sample populations. It sends the message that dealing with continents, race, or ethnicity is arbitrary.[20]

Serre and Pääbo's efforts attempt to simultaneously solve society's race problem while closing an earlier debate over whether starting with well-defined populations leads to sampling biases and thus the production of faulty taxonomies. Are scientists misrepresenting race as a result of sampling biases? What does *race* have to say about the *genome*? Pääbo argues that studies on Americans lead genomics down the wrong track, because self-defined American racial groups are an unrepresentative sample of global or species diversity. As an alternative, Pääbo says, "Rather than thinking about 'populations,' 'ethnicities' or 'races,' a more constructive way to think about human genetic variation is to consider the genome of any particular individual as a mosaic of haplotype blocks."[21] Pääbo explains that each haplotype has a unique history that is shared among individuals, groups, and even genera and species. He asks scientific audiences to beware how studies conducted on well-defined populations are interpreted.

Luminaries Richard Lewontin, Mary-Claire King, and Marc Feldman have made statements that support these efforts. They have agreed that though race, in terms of continental clusters, can been deduced from the genome on the basis of select markers, these markers are "not typical of the human genome in general."[22] Therefore, basing medical analysis on such findings will lead medical professionals to miss the normal transspecies variation that underpins disease. Pompeu Fabra University's Francesc Calafell has added that no matter which groups are sampled, well-defined or admixed, biases are always present in genomics. Pointing to the HapMap Consortium's sample of three distant populations, he encourages scientists to be wary of these increasingly systemic biases.[23] This argument that races can be discerned but are medically moot is a far cry from Venter's White House remarks that ethnicity can't be ascertained from the human genome. Genomicists have moved away from the simple message that the genome proves race does or does not exist, and toward using the paradoxes of race to better understand the science in general.

Pennsylvania State University's Ken Weiss and Rick Kittles have summed up these reflections on sampling biases: "careful geneticists know that if we had more geographically comprehensive samples, human genetic variation is actually characterized by clines (spatial gradients) of allele frequency rather than categorical variation between populations, and the pattern varies among genes for the historic reasons of drift, selection, and demographic history."[24] Kittles and indigenous neurogenetics expert Jeffrey Long have also argued that the most basic population coefficient in the toolkit of all modern geneticists, the F statistic, is itself subject to bias.[25] In another genomic measure of race, they have found that comparing intergroup differences between eight human populations and several chimpanzee populations produces a figure that approximates the same 15 percent figure, no matter what groups they choose. Long and Kittles conclude, "The patterns of variation within and between groups are too intricate to be reduced to a single summary measure. . . . It is now time for geneticists and anthropologists to stop worrying about what does not exist and to discover what does exist."

In questioning race, genomicists have shown that they are willing to put their fundamental tools into question, yet in the interests of calibrating new taxonomy. As Venter and his institute's former project director of human genetics, Susanne Haga, have argued, gene targets are better markers of human difference than any other. In a policy statement, Venter and Haga have warned genomicists against importing Census categories into pharmacogenomics. Contrary to others who continually invoke the imminence of personalized medicine, and thus the need for a proxy in the meantime, they say individualized medicine has already arrived. Indexing gene-targeting diagnostics entering the clinic, Venter and Haga have directed clinicians and scientists to focus on gene targets, not race.[26] Goldstein has sardonically concurred, "Only in the ignorance of [genetic information] do you think about the population and flop on a racial label and say, 'That's good enough.'" Ken Kidd has agreed that the "structure" present in the genome is "extremely small," and thus there is no genomic basis for race.[27]

Even these challenges have been turned into opportunities to reinforce the image of genomics as "rationalized medicine" and the only true source of racial expertise. In Haga and Venter's policy recommendations, they have asked the FDA to reevaluate its draft guidance on the basis of genomic data, implying that the government needs to catch up to science. They also recommend that funding agencies allocate more resources for genomic research into race:

> If distinct genetic variation between racial and/or ethnic groups could be identified, a second tier of studies would be required to determine whether such variation correlates with diversity of disease risk or drug response. If race is to be used as a proxy for nongenetic variables (such as level of income or education), there will need to be studies examining how social factors influence disease and drug response, and whether these factors can be quantitatively measured.

With this three-step agenda, Haga and Venter (2003) argue for genomics to take over what medicine has yet to get right. As Venter has put it, "Many in the medical research community simply don't get it. . . . [O]nly minor statistical differences occur . . . yet doctors may think there's some scientific rationale to give African Americans different drugs than Caucasians because of minor statistical differences."[28] Venter and Haga even suggest that social policy *beyond* the clinic be rethought in light of genomic findings about race. Haga has aimed critical remarks at federal racial enumeration: "It would be inaccurate to check off any one box on the U.S. census if you are African American or Caucasian because to some degree we are all admixed."[29] The editors of *Scientific American* seem to agree. Reflecting on California's vote to maintain racial enumeration in state government business, the editors have retorted that genomics has already proved that genetic tests can replace race.[30] Though they acknowledge the offsetting expense of constructing genetic profiles for each and every citizen, they advise readers to be wary of health findings based on race.

Since midway into the decade of the genome, genomic technologies have been solidly at the center of the problematic of defining race, and the realities of the genome have been accepted as the final arbiter of race. As genomicists have gained public notoriety for their studies of race, it has become normal to use genomic technologies to measure race. Scientists no longer tread lightly around the subject with once-favored terms like "ethnicity" and "self-identified groups." Rather, they openly highlight the problem of *race* in variation work. Scientists on all sides of the debate assume the authority of genomics.

A defining statement from the field is *Nature Genetics'* "Genetics for the Human Race," a compendium of comments from Howard University's 2003 conference. This supplemental issue consists of eleven articles publicizing an array of views and topics that continue to be the key aporiae scientists struggle over. In the sponsor's foreword, Ari Patrinos of the Department of Energy displays just how invested the government has been in building genomic expertise on race. Patrinos frames genomic inquiry into race as a public mission. He describes the participating scientists as stewards of the public good—"entrusted with public funds" yet already making good on their promise with the "solid ground [of the genome sequence] from which to move forward." The editors of *Nature Genetics* agree that public awareness of genomic findings is sorely needed: "Unless we discuss human genomic variation, we will all miss the emerging story of who we are and where we come from." Though they claim that genomics will never solve the larger problem of *racism*, the editors acknowledge that science is an original source of racist ideas and thus shoulders the responsibility to do as much as possible to correct public misunderstandings. They call on genomics to turn the tables by promoting genomic information in place of unscientific notions of race.[31]

In the first of three commentaries, Georgia Dunston and bioethicist Charmaine Royal call the idea of race a paradigm that needs to be shifted.[32] Yet similar to other *race*-first positions, theirs also outlines the need for research specific to "African Diaspora populations." They maintain that their translational genomics research in the African diaspora project will be a first step in the direction of a genomics-led, gene-environment framework for solving health disparities.

Francis Collins lists seven strides urgently needed. First, he calls for genomicists to move beyond racial taxonomy. Next, he calls for scientists to carry out "well-designed, large-scale" gene-environment studies that can provide so-cial epidemiological data as well as genomic data. He also specifies that future genomic research should enroll multiple groups for longitudinal and epide-miological studies and asks for continued support for global survey studies that will cover variation pertinent to minorities. Collins suggests that social science research on identity and representational issues is sorely needed and thus calls for surveys of genomic uses of race and ethnicity. Finally, he directs scientists to educate "researchers, health-care professionals, and the general public" about race and genetics. Collins's comments confirm that the field's leadership be-lieves most of this work needs to be carried out by genomicists. Only one of his recommendations enlists expertise outside the field.[33]

In seven perspective pieces, some of the field's top scientists give their own review of the field's present position with race. The first article, by Joanna Mountain and Neil Risch, claims that genomics has not yet determined whether race is in fact genetic. Yet as in prior arguments, they remind readers that "ge-netic" and "biological" are overlapping but different qualities. Contrary to the rest of the panel, Mountain and Risch argue that race *is* relevant to biological processes, even if it is a poor proxy for genetic variation: "Racial or ethnic cat-egorization will continue to be useful as long as such categorization 'explains' variation unexplained by other factors."[34] Kittles, Shriver, and Parra convey the problems in using race as a proxy for genetic variation by showing that skin color does not always match genetic ancestry. Genotyping and measuring melanin content in "five geographically and culturally defined populations of mixed ancestry," they have found that skin color and ancestry are variably cor-related and thus variably predictive among different ethnic populations.[35]

Jorde and Wooding present their *Alu* data and provide a review of phar-macogenomics. They maintain that geographic origins are a better proxy than race for genetic ancestry. With Collins, they celebrate the field's ability to dispel unscientific beliefs about race: "Fortunately, modern humans can deliver the salutary message that human populations share most of their genetic variation and that there is no scientific support for the concept that human populations are discrete, non-overlapping entities." Charles Rotimi clarifies the purpose of population studies like the HapMap Project and Howard University's African biobank: the studies' populations were not meant to be generalized and thus should be a beacon for future research. He talks about his earlier Howard bio-bank project, another experiment to illustrate how to elicit *within*-population differences by studying gene-environment interactions among various populations of African descent. He holds the press and the pharmaceutical industry responsible for twisting these measures into a reification of race.[36] In a subsequent article, Rotimi, Kittles, Dunston, and their colleagues at the National Human Genome Center condemn the typological fallacies of the idea of race.[37] They argue that humans do not meet the zoological criteria of race as subspecies, again deferring to science to decide whether race is valid.

The final articles in this issue touch on the sensitive issue of pharmacogenomics and population-specific alleles. Goldstein and University College colleague Sarah Tate caution that genomics could cause health disparities if it ignores minority health. Citing twenty-nine pharmaceutical therapies that by 2004 claimed different racial effects, they argue for race to be taken into account until more is known about genetic and environmental causes. Their solution: "more and better research in those groups that have been traditionally underrepresented."[38] In this instance, inclusivity and fears over health disparities trump acuity. Goldstein and Tate argue closer to Risch's position that a meantime solution is needed. Contrarily, Kidd and his former student Sarah Tishkoff close by noting the preference genomicists have for direct markers of genetic ancestry, but also the existence of specific health-related population profiles delineated by immunological and metabolic processes. They argue a hard "antirace" line and urge scientists to treat individuals as opposed to populations.[39]

Since the appearance of these statements, there have been numerous repeated attempts to use sample comparisons and other techniques, such as personal haplotype mapping and whole genome sequencing, to create a new picture of human variation. Some of this research has ventured into sociogenomic waters quite far from the core of molecular science. Two notable efforts include the personal genomics company 23andMe's forays into university epidemiology studies, and DNAPrint Genomics' social epidemiology software development. Since 2009, 23andMe has partnered with scientists at Stanford University, the University of California, Berkeley, and Morehouse College to offer students nonmedical health and ancestry tests. The Berkeley project, "Bring Your Genes to Cal," set out to genotype all incoming freshmen at loci that confer information about alcohol, folate, and lactose metabolism. After a heated public debate about the ethics of sampling students without genetic counseling or physician involvement, and the legality of using such data for academic research, the Department of Public Health ruled that neither Berkeley nor 23andMe could report information on or to individuals. The more successful, race-focused Morehouse project was led by Mark Shriver, who used DNA and three-dimensional photos collected from 150 African American men to raise awareness for personal genomics, while furthering forensic research into the genomics of craniofacial structure.[40] This project fell in line with Shriver's prior dissertation research, which produced the first racial ancestry informative marker technology used in forensic science, and later craniofacial research that established the DNAWitness technology mentioned earlier.[41] For this project, Shriver has sought to understand the relationship between four levels of ancestry: genealogical (what is known through family history), genomic (a person's genomic profile as understood through the sequencing of ancestry informative markers), phenotypic (a composite made from the measurement of expressed traits like skin reflectance and nasal width-to-length ratio), and ascribed (how others perceive and categorize a person). One component of Shriver's research questions how accurate people are at perceiving others' genomic ancestry. The Berkeley and Morehouse projects thus marry DNA research with research into subjective processes linked with racial identity. All the while, they foster a stronger relationship between industry and academia, and between natural and social science.

Perhaps an even more unexpected example is the development of industrial bioinformatic platforms for social research and policymaking. In an attempt to put their technology on the side of a critical science of race, DNAPrint Genomics eventually teamed up with a software company to create a social epidemiology tool. This product garnered a sum of funding from the NIH that was much larger than the ceiling for noninstitutional social research grants across government agencies.[42] The software company's press secretary explained the race-positive intent of the software:

> The need within American society for new research on the topic of race is large and growing. And the need among demographers and other social and health scientists for additional research tools allowing them to carefully explore the social and cultural elements surrounding racial identities is equally great. The recent directive by the federal Office of Management and Budget requiring the option of multiracial classification on federal forms; the growing racial and ethnic diversity within American society, coupled with the knowledge that racial identities can vary depending on measure, context, and cultural factors, are evidence of the need for additional tools to help clarify the circumstances surrounding the shaping of racial identities.[43]

Here genomic technology is being called upon to address the social conundrum of race, to eventually create policy recommendations to correct federal norms.

The labs surveyed here continue to make new contributions to the debate over taxonomy. The iterative nature of specific labs' participation conveys both the gravity of the race problem for the field of genomics and the interpellatory power the debate has across the field's elite sector. Race is under study by all—a normative effect of the constant struggle to produce a positive relationship with the public over matters of difference and a new way of imagining it.

It is fair to say that one of the common dilemmas by which the field builds its reputation is this one of race. It is not a matter of isolated protocols but rather a field's activism to solve a defining problem of modernity that characterizes genomic science. Genomic expertise about race involves constant reflection on what new tools and data have to say. Yet far from engendering uncertainty in the field, these disputes produce a solidarity born of common cause and experience. As a collective, genomics invests in its role as arbiter of race through a proactive, race-positive inquiry.

Interdisciplinarity and the Status of Genomic Expertise

As with lay activism organized around a central principle yet full of internal struggles, genomic activism is able to forge contentious coalitions across the domains of racial expertise at the same time that it constructs a unique core identity for the field of genomics. A distinctive mark of the field is the collaborations it has made with social scientists and humanists in recent years. These collaborations take the form of conferences, seminars, and publications that commit the field to a sociogenomic optic and to proactively revisiting race with each new technological advance.

Take for example Stanford University's 2003–05 interdisciplinary seminar sponsored by the Stanford Humanities Center, Affymetrix Corporation, the Mellon Foundation, and the Research Institute of the Center for Comparative Studies on Race and Ethnicity—a series which led to a book and journal articles on race and genomics, coauthored by scientists and critics. Excerpts of the seminar's 2009 statement on race, published in *Genome Biology*, shows that scientists' predominant vision for going forward in the field is through a collaborative, interdisciplinary approach:

> Statement 1: We believe that there is no scientific basis for any claim that the pattern of human genetic variation supports hierarchically ranked categories of race or ethnicity. Furthermore, we abhor any use of genetic data to reinforce

the idea of between-group difference in order to benefit one group to the detriment of another.

Statement 8: We encourage the funding of interdisciplinary study of human genetic variation that includes a broad range of experts in the social sciences, humanities and natural sciences.

Statement 10: We recommend that the teaching of genetics include historical and social scientific information on past uses of science to promote racism as well as the potential impact of future policies; we encourage increased funding for the development of such teaching materials and programs for secondary and undergraduate education.[44]

Though many nonscientist panelists with whom I spoke reported feeling frustrated and barely listened to at many points along the way, none denied that genomics has worked consistently toward an interdisciplinary investigation of race.[45] It is through these interdisciplinary panels, spawned in the mold of ELSI, that the field's main policy around race is formed and scientists are forced to consider alternatives to their intuitive practices.

Another notable instance of collaboration has been the Ancestry Testing Task Force of the American Society of Human Genetics. The task force emerged from the society's Social Issues Committee, an interdisciplinary group that brings in social scientists like Stanford University's Sandra Soo-Jin Lee, legal scholars like the National Human Genome Research Institute's Vence Bonham, and bioethicists like Duke University's Charmaine Royal. The committee began working on the ethics of ancestry tests middecade with a panel of elite scientists, including several past presidents of the American Society of Human Genetics.[46] This led to two official statements on direct-to-consumer testing[47] and a 2010 annual meeting panel, presided over by Lynn Jorde, entitled "Researchers Shed Light on Implications and Impact of Using Direct-to-Consumer and Clinical Genetic Testing in Disease Risk Assessment."

The Ancestry Testing Task Force is representative of the breadth and caliber of policy partnerships in the field—racial expertise is formulated in interdisciplinary committees at the world's leading genomics centers. It is also representative of a structural feature of the field, mentioned earlier in the book: in-house social science, epidemiology, and policy research branches. The task force's 2010 statement, "Inferring Genetic Ancestry: Opportunities, Challenges, and Implications,"[48] was coauthored by affiliates of such hybridized institutions as the

Institute for Genome Sciences and Policy at Duke University, and the Center for Society and Genetics at the University of California, Los Angeles. The mission of these centers is, in the words of the Center for Society and Genetics,

> to educate the public to understand, discuss, and make informed decisions about issues in biotechnology, genetics and genomics. Important to the CSG's research and educational endeavors is the focus on diversifying the community of students and faculty. Diversity encourages a broader scope of research that will inform scholarship with insight from, and relevance to, diverse cultures. Genetic research and discussions surrounding technological developments will be enriched by, and will co-evolve with, the cultural diversity of the voices in those discussions.

Here diversity is defined within the *academy*. Yet this trope dovetails with the burgeoning focus on racial diversity in the labor force and in research sampling.[49] Duke University also states: "We are dedicated to the study of life through scientific inquiry involving interdisciplinary research in genome sciences and policy. We are a passionate supporter and facilitator of campus-wide research and scholarship that explore the impact of genome sciences on all aspects of life, human health and social policy." These institutes make interdisciplinary dialogue and work inherent in all aspects of the research process. They not only provide a space for dialogue but ensure collaboration across projects. Leadership of these elite genomic research centers is provided by genomicists *and* social scientists as well as humanists. Meanwhile, public health departments such as the NIH and the U.K. Medical Research Council provide monetary incentives to institutions that prioritize interdisciplinarity.

This interdisciplinary foundation shapes the way genomicists pursue inquiry into race. Formal and informal probing of the social and political implications of such inquiries makes it easier to speak about the political, and also personal, stakes in genomic matters. Constant demands for scientists to own up to their biases train them to think like social scientists and humanists, and to become comfortable advocating for certain positions. In these settings, acknowledging one's partiality is highly valued.

One might then wonder how it is possible for the field to promote a sociogenomics of race while it pushes for a genomic overhaul of race. How can a science intent on providing a new taxonomy truly be invested in the social questions that motivate alternative sources of expertise on race? This contradiction points to a hierarchy within the interdisciplinary enterprise and within genomic thought. Society values natural scientific solutions to social questions just as genomicists privilege genetic explanations in the last instance. Though multiple streams of interest shape how genomics revamps race, the patently biological DNA code remains the Rosetta stone of human diversity.

Here Pierre Bourdieu's discussion of fields and capitals is a useful lens with which to understand why ultimate allegiance to a natural scientific answer is so important to genomics.[50] Social fields such as the economy, family, science, and culture are sites of conflict in which members of dominant and subordinate classes struggle for, amass, and leverage different forms of capital. Yet the scientific field is unique in that science, with its charter for describing reality, holds a special claim over reason. In Bourdieu's terms, the field's capital generates symbolic currency with all of society. While there are fascinating ways that this shapes the internal struggles of science, such as restricting dissension to peer-reviewed journals and academic conferences, what is of interest to us is that genomics must maintain its claim to authority over nature, in all instances. At issue is billions of dollars in resources, but also license to guide the public's relationship to the world around it.

Any rough comparison of funding streams in the natural and social sciences and humanities will show that Western societies hold natural scientific forms of expertise in higher regard. Though research costs indeed differ across fields, the order of spending disparities goes far beyond that explanation. For example, in 2009, U.S. university spending in research and development totaled $32.8 billion for the life sciences and only $2 billion for the social sciences.[51] Similarly, the National Science Foundation spent $3.7 billion on science and engineering research and only $200 million on social science and psychology research. The biomedical giant Johns Hopkins University received the greatest amount of research and development funds of any academic institution—gaining almost double the amount of the next-ranked "science and engineering" school and *twenty-six times* the amount of the top non–"science and engineering" institution. In fact, the U.S. government prioritized biomedical research in *its* funding; the Department of Health and Human Services, which houses the NIH, came out as the leading funding agency for domestic academic institutions.

This financing disparity does not detract from scientific interest in interdisciplinary but rather gives interdisciplinarity a symbolic value to scientists. First, by maintaining a connection to the larger debate on race—whether disputing solutions, providing new visions of taxonomy, or permitting the influx of racial classification by self-identification—scientists expand their public image and social relevance. Second, by engaging in the controversy and becoming familiar with the terms of other racial experts, scientists are able to manage the controversy while appropriating alternative views. Third, through the give-and-take of interdisciplinary dialogue, genomics fabricates its identity relationally to forms of expertise that don't have the capacity to master the biological side of critique. One of the field's central organizing points is its ability to define itself as the authority on humanity, on multiple levels; thus, its sociogenomic efforts remain key to the legitimacy of its knowledge base.[52]

Above all, genomicists fashion themselves as agents in a biosocial world that they themselves are rendering. It is no accident that the relationship between biology and identity has become a pressing question for society just at the moment when biomedical knowledge is transforming our foundational social institutions. The market in drugs, cosmetic surgery, nutrition, and direct-to-consumer genomics forms a baseline against which members of society increasingly understand and transform themselves. In the words of Adele Clarke, and a number of other scholars of the contemporary, society is becoming increasingly "biomedicalized."[53] In addition to disease, we find that health, emotions, habits, and everyday life processes have become objects to be diagnosed, managed, and treated with medical cures. New informatic systems, clinical innovations, and a burgeoning biotech sector expand biomedicine's reach and form the context for the biosociality genomicists' advance. Because people are expected to engage with bio-literature, manage their health, improve and enhance their bodies, and prevent illness and decay, these structural transformations in medicine and healthcare ensure that a genomic wash for "life itself" will be salient for constructing identity and subjectivity.[54]

We are witnessing a change in the capacity for humans to know and transform their biology; the pressure to think and act "bio" is compelling. Yet this is more than a sea change in opportunities for individuals to optimize their bodies and customize nature. As Rabinow has argued, nature never has been free of human input, at least as far as it is relevant to humans; similarly, there never has been a body without artifice. But now society has moved from using bio-metaphors for social life to taking biology as the *principle for social organization*. Still Rabinow's interpretation also falls short, because it fails to recognize how older classification systems are operative, entwined in new taxonomies. Genomicists struggle to produce a new taxonomy because they desire *a new order of groupness*. This is not the same as wanting to provide opportunities or take up opportunities to transform the individual body. Rather, it is about the establishment of tools for the maintenance of social positions and relations that people hold dear.[55]

Genomicists put themselves in the position of being advocates, activists, and activators of a new world in order to move that world into a new intersubjective framework. Inasmuch as race remains a fundamental principle of social organization, their prototypical biosocial science will continue to make group identity and group solidarity a research priority. Scientists attempting to change the status quo cannot extricate themselves from its basic principles, because these lie at the personal and collective levels. As they attempt to produce a superior knowledge about human diversity, genomicists reformulate rather than eliminate notions of the clan and the collective. Not only addressing the field of possibilities individuals face, they alter the way groups will collectively see themselves.

Biosociality as a Way of Life

In an op-ed on race-based medicine written for the *Boston Globe*, Goldstein and Duke University's Huntington Willard called on the memory of Dr. Martin Luther King Jr.: "as we recall how King compelled us to reconsider how race affects education, employment, and social justice, we should sharpen our focus on new racial challenges posed by emerging medical advances."[56] Their remarks, which use the language of minority justice and hold a tenor of social activism, show that genomics is investing in a specific biosocial version of social equality to be delivered by a specific group of experts.

Though the field is divided over how race should be conceptualized, it is resoundingly committed to being *the* new science of race. Scientists on all points of the spectrum believe that genomics should be entrusted to define the ins and outs of race. Most also believe that genomic evidence should inform policymaking. The field continues to test race in the belief that better knowledge of the genome will produce better social relations. Scientists hope that the truth of the genome will be society's racial truth.

In recent years, scientists have grown optimistic about their ability to set the racial course for biomedicine and society. As with the personalization of the genomics of race, a confidence in nonscientific values promotes a broader belief that genomic science is sound. While scientists incorporate social constructionism into their rubric of knowledge about race, they assert their taxonomies as the ultimate truth. This acceptance of responsibility for public understanding thus reinforces the trend toward using genomic metaphors for race. Genomic visions of the future show that the sociogenomic paradigm is fueling a deep belief in biosociality as a way of life. Yet as scientists attempt to create new concepts of the human and new lenses for relatedness, older principles of difference continue to be defining dilemmas of the field.

What does it mean for genomics to be the last stop in racial science? How far can genomic activism take us? Though genomicists convey a deep desire to improve the social landscape, there are signs that genomic activism has serious limits. For all of their talk of thought-leading and antiracism, elite scientists have often done nothing more than verbally critique overtly racist representations of genomic data.

In one instance, a team of researchers at Howard Hughes Medical Institute led by the University of Chicago's Bruce Lahn reported the positive natural selection of brain size expansion, the hallmark of human intelligence, in non-African populations.[57] Lahn and his collaborators claimed that brain size expansion was correlated with cultural innovations only occurring outside of Africa.[58] This research was released amid a flurry of new findings on positively selected traits like lactose tolerance and resistance to malaria,[59] so though it was debated, it consequently received a great deal of attention and respect within the science community.[60] Since then, but for two teams of scientists, the field has not come out against the findings. Even Lahn's elite collaborators, some of whom feel abused by his misuse of their samples, have not taken a stand.

Similarly, in the case of BiDil, many scientists have gone on record professing criticism, but none have gone so far as to challenge BiDil from a policy perspective. Ironically, it has taken one of the original BiDil researchers, cardiologist Clyde Yancy, to give an in-depth public denouncement of BiDil. He has argued that the American populace is far too admixed for race to be used as a guide for prescriptions, and has emphatically said: "We need to move away from race quickly. As we mature, we will be able to supplant the notion of race as predictor of response with something more palatable to the scientific community and to patients. Then we don't have to bring the word heft of 'race' into how best to care for patients."[61] Keith Ferdinand of the Association of Black Cardiologists has meanwhile come around to criticize the racialization of the therapy, arguing that whites should get the drug too.[62] The weak response by genomic elites to clinical uses of race suggests that genomic activism stops short at research politics. Despite claims to expertise and moral authority, scientists are choosing to ignore what happens outside their immediate realm.

Trends in biosociality and biomedicalization show us that the world's vision for an ultimate solution found in our DNA may take us further away from the social inequalities that the field purports to address. Because genomicists begin with the "genomic" rather than the "socio," the biological effects of structural racism remain a secondary consideration. Social factors are pursued as aftereffects of genetic activity. The "socio" is often a restricted metaphor for individual habits, whether personal health choices or researcher communication strategies. This narrow framing of the "socio" does not provoke scientists to consider the historical and present-day racialization of groups, nor does it capture the identity processes central to the lives of racial minorities. The dialogues underpinning racial consciousness-raising in the field suggest that scientists will continue to experiment with models for understanding race. However, asymmetries between the status of biological and social research will likely encourage a further emphasis on biological solutions. As I quoted Jonathan Kaplan earlier, "biology doesn't make 'race.' But 'race' might very well make biology." In order for scientists to deliver a biosocial world where justice reigns, taxonomy must be not only remade but rethought. Exploring where the field is headed and how the context for understanding equality is changing will further shed light on these issues.

7 The Enduring Trouble with Race

I N LITTLE MORE THAN A DECADE, genomics abruptly about-faced into a proactive, race-positive science. As we have seen, in the early years of the Human Genome Project, genomics largely ignored race. Researchers across the field deemed race a social matter, a clinical hazard, and a scientific fallacy that genomics had little to do with. Though the field responded rapidly to accommodate the policy shifts that commenced in the mid-1990s, conceptual and practical adjustments were less instantaneous. It was not until the publication of the human genome that scientists began making authoritative statements about the social meaning of race and seeing the field as a major contributor to public health. Similarly, at the dawn of the government's policy break toward health disparities research, most scientists were still responding to ethical quandaries like individual privacy and education about genetic risk management. Yet by middecade scientists had embraced the role of arbiter of racial knowledge. Though contention rose as to how scientists should implement inclusionary policy and a minority justice ethic, the field was unified in its approach to race as a problem for genomics to solve. Scientists with a variety of explanations for race, motivated by a keen interest in minority justice, made getting race "right" a central aim of genomics, thus bringing the field to where we are today.

This shift has depended on the transfer of racial paradigms between the broader society and genomic expertise. Genomics moved away from colorblind science just as colorblind politics became untenable in the general public. The race-positive paradigm that was first promoted in Congress and in the field's main funding agencies made its way into the Triple Helix of science via the ethical, legal, and social branches of the field's major institutions and projects. This racial paradigm has led to the fortification of a racialized system of databases and a racialized sampling protocol across the field. Genomics is part of a wider social shift but is also authorizing and driving it.

The rise of genomic racial expertise in the public domain and the gradual amplification of contention amidst ELSI ethicists have also been transformative factors in how the field thinks about race. In battling for particular positions and ideas, scientists have become public figures. Participation in the growing controversy has interpellated genome scientists, quickening a reflexive, personalized scientific ethos bent on biosocial solutions for the world. Genomic elites have pioneered a sociogenomic consensus in which race is something social and biological, requiring subjective and objective frames of analysis. What genomicists call race now is equivalent, but not identical, to what they called race a decade ago. Groupings based on ancestry, derived from clades, and depending on mitochondrial DNA or Y-chromosomal mutations are widely agreed to correspond to old racial categories, though they are now understood to be the result of both social and biological processes. This new paradigm is so effective because it can be used for many of the same causes that the old one was used for—such as minority inclusion and racial justice politics. Inflected with the goals of social empowerment and diversity, the sociogenomics of race allows researchers to communicate with the public and with critics in a familiar idiom. This creates an ambivalent political field in which minority justice and identity politics grows dependent on genomic classification, processes that amplify the field's position across biomedical institutions and markets. Inattention to the structural violence inherent in the production of race, and failure to significantly engage alternate knowledge frameworks, are but two important consequences, which I further address below.

An even more complicated picture arises when we examine the ways genomicists think and work with race in the research domain. A dynamic conceptualization of race is at play, where multiple working models with varying social implications hang together. This provides scientists with a flexibility, if not an agility, in making definitions, which is useful for enrolling different kinds of support. As a result, a values-based pragmatism has sedimented into a central principle of the conceptualization process. Scientists overwhelmingly oppose standardization, or any conceptual closure, in the field.

An equally dynamic scientific practice predominates. On the one hand, an infusion of identity politics creates multiple frameworks of reflexivity, and promotes a measure of researcher autonomy in the preparation of protocols. On the other hand, as scientists attempt to put race-positive values into effect, they collectively sponsor concessions to social taxonomy. Thus, rather than make value-pragmatism a cowboy affair, researcher autonomy leads to shared sampling strategies and common frameworks for reporting genomic findings.

An activist ethos spurs the field's notoriety as the be-all and end-all science of race. Genomicists' confidence in the field's ability to challenge past racist notions with new taxonomies drives them to iteratively test the genomic validity of social descriptors. Though they build their public persona on interdisciplinary foundations, elite members of the field and their respective institutions still rely on their image as the most rational science of all. The field's expertise takes the shape of a kind of biosocial advocacy, where the notion of sociogenomic race is reified at the same time that genomic taxonomies are sold as the solution to contemporary social ills.

Among several key sociological implications of this chain of events, three stand out in creating productive dialogues about science and human difference going forward. First, we must begin to recognize that identity politics not only is a property of the mainstream political arena but is inherent in science as well. Though genomicists operate within a larger politics of classification, including that of federal policy and global pharmaceutical ethics, their personal experiences and beliefs inform their perception and practice. Their own identity politics conditions their approaches to race. Recursively, that politics creates a certain field of possibilities for lay identity construction. In other words, scientific identity politics loops back into the identity politics of policymakers and the public.

Second, though genomic taxonomies are just as likely as older taxonomies to reify social differences, we must understand that genomicists' attempts to solve racial problems with biological solutions stem from the same historically shaped dream of a just world that motivates other racial justice activists. Those of privileged backgrounds too take a hand in the prevailing racial paradigm. Genomicists use their privileged positions and the tools they know best to produce an antiracist racialism. Race, conceived as encoded in biology, is not a morally neutral stance but rather an active attempt to build a new moral order.

Third, scientists are cognizant of their identity politics and support the mobilization of personal value-pragmatism in research. Minority scientists reflexively fashion themselves as exemplars of embodied equality, while scientists who identify as white or Jewish model how to witness injustice and support the "living *as*" of others. Asking scientists to take heed of the subjective side of knowledge production is thus anachronistic and does little to bring a fuller social understanding about. It would be more productive to focus on the values that are being mobilized in research projects. What assumptions do scientists bring with them to the table? How do scientists' beliefs link up with ideas at work in racial projects that are external to science? Which values should be encouraged and made to count?

To understand where we might go from here, I discuss some of the limitations and ramifications of the American-dominated sociogenomic paradigm. I use this analysis to reflect on the drawbacks of authorizing narrow forms of expertise to single-handedly solve deep social problems. First, I consider the questions that the globalization of genomics presents as more stakeholders are unevenly drawn into genomics. I also take a look at the dilemmas inherent in the field's move toward behavioral science. Finally, I discuss the racial paradigm that is just on the horizon and offer some clues as to how we can promote a more socially responsible science of human difference. These issues pertain not only to the social production of race but to authoritative knowledge writ large. They illuminate the perpetually precarious dynamics of expertise as scientists remake the world we live in, and the possibilities of how life can be lived in such a world.

The Asymmetries of a Transnational Science

At the end of the decade of the genome, both policy and protocols are rapidly globalizing. The U.S. federal government, one of the largest sources of funding for genomic research worldwide, is expanding into new markets and territories. Just as biological citizenship and "bioconstitutionalist"[1] forms of social struggle are taking shape—where political struggles formerly fought in a social idiom are now biologized with demands for the individual body—we see the emergence of a new human rights–based model for international research and a humanitarian biosociality. Undertakings like the Genographic Project and the Human Health and Heredity in Africa Project claim to create benefits for the people sampled and to extend the fruits of genomics to developing nations. "Helicopter genetics," or research done on subjects in a "peripheral site" to meet the needs of the West, is the foil of these undertakings. Projects build their public persona around the notion of community investment and expansion of the knowledge economy. These promises show that racial science is evolving within a context of *global inequality*, where the term "uneven playing field" holds new meaning for scientists.

More than a globalization of global research, this is the transnationalization of transnational research. Research is to be constructed on site, by local teams of researchers. Even projects governed from the United States and the United Kingdom claim to be "by and for the people." A fundamental goal is to empower local regimes to be global players. As the Wellcome Trust's Mark Walport has said about the Human Health and Heredity in Africa Project, global genomics isn't just about research: "It's also about people. It's about developing capacity, in order to make the most of human genetics. Now one needs people who are skilled in bio . . . so it's about training people, and about building institutional capacity, the capacity to provide an environment for this sort of research for first-class health research."[2] Walport's appeal joins the ethics of inclusion and empowerment, but for a distinctly transnational moral sphere. The NIH and Wellcome Trust currently fund over fifty African institutions and plan to produce more local genome institutes so that participating centers can become self-sustaining in the coming years. The Genographic Project's Cultural Legacy Fund similarly supports a range of cultural revitalization projects that are themselves selected by a committee of indigenous peoples representatives. Funds are donated to communities for participating and nonparticipating tribes. Thus, inclusion is increasingly being envisioned in a global, material sense.

Yet tensions arise as researchers encounter pushback for the genomic way of defining groupness. In a recent protest against the Genographic Project, Benito Machacca Apaza, the president of an indigenous community in Peru, remarked: "The Q'ero Nation knows its history, its past, present and future is our Inca culture and we don't need any so-called genetic study to know who we are. We are Incas, we always have been and always will be!"[3] The same day, Vanessa Hayes of the J. Craig Venter Institute published a policy piece stating:

> Going forward the science community must do more to secure the participation of indigenous communities in the research process, ranging from correct data interpretation and acknowledgement in science publications, to negotiating potent long-term commercial benefits, to maintaining ties after the research is completed. Access to such communities requires a long-term social obligation that considers basic community needs, such as access to water, nutrition, education, and medicines.[4]

Hayes speaks of the need for sensitivity to the context of environmental abuses and biopiracy, sociological issues like communitarian structures and values, language, and methods of evidence validation. Like Apaza, Hayes says local knowledge should be respected. But in a world where some subjects are seen as potential consumers and others as research subjects-in-waiting,[5] current ways of managing inclusion don't get at the issue of asymmetric power, a framework that ensures that Western science will continue to overwhelmingly define the biosocial world. Cultural competency, and even attention to health access and water resources or education, is moot when such knowledge bases are structured in such a way that one group has more political say.

These tensions are also reflected in nationalist projects, like the Genomic Crusades of Mexico and the Indian Genome Initiative. As governments attempt to construct nationalized pharmaceutical and biotech markets, they make choices about whose DNA to include and make public. These projects frame DNA as a national resource. As Ruha Benjamin has argued, in these contexts, "national diversity is not simply mapped by genomics, but recalibrated vis-à-vis genomic findings in attempts to genetically brand the nation as a niche ethnic market with minimal political fallout."[6] Governments politicize specific constructions of national identity that pivot on marketable notions of ancestry. In the process, some groups within the populace—often indigenous peoples, minorities, or ethnic hybrids—are glossed over, silenced, or ignored.

In addition to national population–based biobanks, governments look to whole genome sequencing to create race-specific models of disease research. Governments seek to brand their DNA as unique in order to gain a toehold in the global marketplace. Korea was one of the first countries to produce a whole genome sequence of a national. Ireland is one of the latest. These projects contribute to the semblance of a "normal" national or "average" racial member.

The expansion of markets and research infrastructures is bringing new publics to the fore that have racialized ways of envisioning and promoting global health justice. Just as minority health organizations have battled for race-specific drugs, researchers of the global market in pharmaceuticals have also demanded a population-based pharmacogenomics for nonindustrialized countries. Worried that the costs of personalized medicine will keep the Global South from gaining life-saving therapies, they have argued that personalized medicine counters the needs of the developing world.[7] Citing a "genomic divide," some advocate for BiDil-esque drugs, interventions that are tailored to national populations in the developing world and that can be made cheap to the public.

As research and development moves to new sites, new value-frames will continue to challenge the reigning sociogenomic model.[8] Local hierarchies, or social distinctions between groups on the ground, will impact how scientists envision and manage difference. We may ask, will research that is conducted away from the traditional centers of genomic power be different? What will we learn from a genomics that is forged in alternate institutional and cultural paradigms? These local achievements are creating new organizational arrangements, conventions, and strategies. They are also being created in the interests of different social politics and classification frameworks. What practices will emerge in this decentralized genomics? What will sampling and labeling look like amid different political landscapes?

How researchers decide to address inequality will necessarily depend on how problems are posed. In a globalizing field, where genomics and its publics are not monolithic, what new ways of cultivating a virtuous science will arise? Whose problems will lead? The scientists presented here show that working outside the United States provokes innovative responses, including altering the mandated inclusionary framework and creating project-specific taxonomies. Yet, racialized policies and markets go undisputed. What will be the value of difference in the coming decade? Will adherence to a majority/minority framework persist, or will new forms of hierarchy come to predominate?

In the case of global projects we may further ask: Will the continental comparative strategy hold up? How will the application of new genomic technologies to new populations under a sociogenomic paradigm transform the way we view difference? In November 2008 the International HapMap Project began releasing drafts of its Phase III dataset, which to the original populations adds people with African ancestry in the U.S. Southwest, people with Mexican ancestry in Los Angeles, Tuscans in Italy, Chinese in metropolitan Denver, Gujarati Indians in Houston, Luhya in Webuye, Kenya, and Masai in Kinyawa, also in Kenya. In September 2010 the National Human Genome Research Institute awarded ten labs over $18 million to continue developing next-generation sequencing technologies with which to study these populations.[9] Yet from studies performed on the 1000 Genomes Project's samples and newly sequenced genomes, we see a new racialized model for studying health disparities emerging. Researchers now start with the disparity and then conduct a whole genome comparative analysis to find different causal factors in each racial group.[10] Some of the latest conditions studied with this method include kidney disease, hypertension, and heart disease. This model uses epidemiological data, in which the balance of biological and social causal factors has not been established, to then seek genetic causal factors for disparities. Social and genomic taxonomies are blurred even further than in the genome-wide analysis of the past decade.

Some believe that the globalization of genomics should be celebrated, as it will expand the realm of the ethically and politically possible and provide benefits to people who are already being enrolled into research. Others argue that enlarging the scope of research can only be done under a program of community self-determination. At this time it is unclear how new stakeholders will be enrolled on equal terms unless the asymmetrical relationship between local value regimes and Western frames is addressed, and sociopolitical hierarchies between regions, nations, and subnational groups are altered.

A Launchpad for the Sociogenomic Future

Behavior genomics is one of the touchiest areas of genomic research and development, and its rapid emergence gestures toward other important challenges on the sociogenomic horizon.[11] Behavior genomics studies neurological processes and behavioral traits using genomic technology. Some of the medical disorders aggressively being investigated are brain cancers, Parkinson's, Alzheimer's, autism, schizophrenia, depression, and bipolar disorder. In addition, cognitive functions related to mental retardation, language, anxiety, addiction, learning, memory, anger, love, and even political attitudes and participation are being evaluated genomically.[12] Animal models have inspired scientists to extrapolate findings to mating decision and behavior,[13] social recognition and behavior,[14] and circadian rhythms.[15] Sociologists and political scientists are also producing oeuvres based on genome-wide association studies of social traits and behaviors like entrepreneurship,[16] leadership,[17] happiness, work ethic,[18] promiscuity,[19] and political persuasion.[20] In behavior genomics we see how racialized genomic discourses recode the "socio" of race into a genetic essence and are thus taken up in ways that lend themselves to racism. These developments provide a cautionary tale for biological sciences intent on studying matters intrinsically defined by social interaction.[21]

Reports on gene prevalence are already alleging the existence of deep racial differences in traits on which society puts a high price. Recall the microcephaly research conducted by Bruce Lahn and colleagues, mentioned in the preceding chapter. This renewed controversy over racial differences in brain development and intelligence illuminates the problems genomicists face as new technologies are applied through old optics like racial divergence theory. Haplotypes, SNPs, and even noncoding regions of DNA are being read through a racialist lens in well-respected labs just as aspects of the social environment are left by the wayside, unaccounted for, and unexplained.[22] Debates and retractions only generate more notoriety for scientists who search for genetic causes. Even after the contention and dismay over racist findings die down, the imprimatur of genomics lingers.

Another example is the growing literature on aggression, in which studies have linked genomic profiles to everything from criminality and "gang behavior" to credit card debt.[23] In 2011, *Science Daily* reported a study confirming a predisposition to aggressive behavior depending on one's racial version of the *MAOA* gene: "Only about a third of people in Western populations have the low-activity form of *MAOA*. By comparison, low-activity *MAOA* has been reported to be much more frequent (approaching two-thirds of people) in some populations that had a history of warfare."[24] Dr. Phil and the National Geographic Society not only featured the research in their programs "Rageaholics" and "Born to Rage" but also publicized where consumers could buy tests for what *Science* has dubbed the "warrior gene."[25] In fact, the original report only briefly cited one population genomics study comparing European and Maori lineages.[26] Nevertheless those findings fueled racial comparisons in a later study that found the aggressive variant in 77 percent of self-identified Chinese subjects, 59 percent of self-identified Africans, 56 percent of subjects with at least one Maori parent, 34 percent of self-identified Caucasians, and 29 percent of self-identified Hispanics.[27] (This study included only *forty-six* research subjects, far too few to be considered robust science.) Studies like these take social hierarchies born of intergenerational inequities as proxies for innate differences. They bolster positions that aim to protect racial inequality. For example, this body of research has since been adopted by North American white supremacist organizations and bloggers in their quest for a scientific backing for their racist ideologies.[28] So far there has been no backlash against the authors of these studies. This reminds us that while genomic elites hold social positions that afford them the chance to "fight back," others in the field may have less resources and incentive to manage the ethical terrain.

Turning social traits and behaviors into genetic disorders poses further problems, because it creates a nurturing environment for other sciences' biologically deterministic interpretations. Magnetic resonance imaging studies on topics like antisocial behavior, drug abuse, and racial perceptions are increasingly touting brain deficiencies and deep biological predispositions.[29] These studies confront research subjects with stimuli designed to capture patterns in brain activity. Some, like the population genomic study above, use sample sizes of fewer than fifty subjects before making species-wide generalizations about the neurogenomic implications of a trait. Like genomicists, they are moved by the promise of solving a fundamental social problem. Yet they operate with even shakier standards of evidence.

Sports biology, too, has taken a famously genomic turn: many researchers and pundits are arguing that genomic interpretations should be extended to athletic aptitude and performance. Jon Entine, for example, the author of *Taboo: Why Black Athletes Dominate Sports and Why We Are Afraid to Talk About It*, uses genomic research to back his claim that people of African descent are genetically preconditioned to athletic events like long-distance running:

> Genetically linked, highly heritable characteristics such as skeletal structure, the distribution of muscle-fiber types, reflex capabilities, metabolic efficiency, lung capacity, and the ability to use energy more efficiently are not evenly distributed among populations and cannot be explained by known environmental factors. . . . The thorny reality is that if there were no "racial" differences, the entire Human Genome Project would be meaningless.[30]

Similarly, speaking on behalf of a team of Duke University scientists who research running ability, Adrian Bejan has said: "Blacks tend to have longer limbs with smaller circumferences, meaning that their centers of gravity are higher compared to whites of the same height. Asians and whites tend to have longer torsos, so their centers of gravity are lower."[31] His collaborator, Edward Jones, agrees: "There is a whole body of evidence showing that there are distinct differences in body types among blacks and whites. Whether the fastest sprinters are Jamaican, African or Canadian, most of them can be traced back [genetically] to Western Africa."[32] These theories merge common racist stereotypes with authentic population genomics. They form a substrate for policies such as the NCAA Division I's mandatory testing for the sickle cell trait. While this policy mandates a genetic test for all of the division's 170,000 athletes, it comes within a tradition of race-targeted policies aimed at limiting African American participation in fields like the military and airline piloting.[33]

Even if scientists were to scale back to only making claims about DNA, there remains a likelihood that behaviors would be compared by race after the racialized nature of global databases like the HapMap and 1000 Genomes projects and domestic databases like the FBI's CODIS system. As Troy Duster argues, research into antisocial and criminal behavior is likely to begin with the mining of forensic databases.[34] Indeed, in California the state government and the University of California system have been in talks about an academic takeover of the California Correctional System's health and medicine program.[35] With a racialized database produced by a discriminatory criminal justice system, racial findings will be inevitable. Meanwhile, as research into noncriminal traits arises from the databases of personal genomics companies and pharmaceutical trials, comparisons will run along familiar stereotypic lines.[36] This comes at a time when large-scale international genome projects and national databanks are pushing for wholly centralized medical data.[37]

How will data from behavior genomics be marshaled in policy? So far, we have seen isolated legal incidences, such as in an American case where a judge predicted recidivism for a child pornographer on the basis of his genotype,[38] and an Italian case where a forensic psychiatrist requested leniency when confirming a killer's *MAOA* variant to be the aggressive type.[39] Yet new population-based studies continue to emerge and brandish criminal behavior as genomic and racial, just as states are adopting ancestry tests to determine the ancestry of political asylum-seekers and immigrants.[40] In 2009 the British government's U.K. Border Agency launched a pilot test of the Human Provenance Project, which was designed to use mitochondrial, Y-chromosomal, and SNP testing to determine the national origins of asylum-seekers who have failed a linguistic test.[41] Little more than a year later, the U.S. Department of Homeland Security announced plans to test the use of DNA scanners at airport security lines.[42] Behavioral findings set in racial or national terms are more important than ever. Deterministic claims about gender, sexual orientation, and other axes of social division may soon follow as social interaction gets further coded as genetic.[43]

Postracial Imaginaries

Today what counts for racism is again changing. As Matthew Hughey argues, in the wake of the Obama elections a Horatio Alger narrative has ascended and contributed to the belief that talking about contemporary racism is passé, if not downright racist.[44] Acknowledging and celebrating differences is just as important as it was at the height of the past decade; however, the mechanics of present-day racial discrimination are an unpopular topic in the political mainstream. In this climate, scientists are under an even greater pressure to find a taxonomic alternative that focuses on past paradigms as a foil. When I first began speaking with genomic elites, talk of replacing race with ancestry was brewing. Now it is a must that ancestry be referred to, even in the most race-positive communiqués.[45] Again we see the hints of a desire for a world beyond race. Antiracist *post*racialism, a biosocial framework for attacking racial disparities, may be the new racial paradigm that will frame future social relations.

Postracialism is typically described as the political retreat from matters of race. In the past it has stemmed from the belief that societies are no longer organized into white supremacist states informed by a "racial contract" that privileges a hierarchy of races. It has also stemmed from the desire to end "living *as*" as we know it. Today many who invoke postracialism indeed believe that society has moved beyond clear structures of racial hierarchy, and that appearances don't affect identity construction and perceptions as much as in the past. Also, in the United States a specific brand of "racial fatigue" predominates.[46] Many Americans just plain don't want to talk about race anymore, because they so badly desire to already exist in a state of equality.

Genomics is reformulating racialism in the lab and field with ethical frameworks inspired by humanitarianism, yet its antiracist approach is still trained on past fields and their hand in racism rather than on the historical and sociological causes of disparities. Global crises, along with other motifs of humanitarian emergency, will likely be increasingly invoked in this vein. With traces of the postracialism, will the incitement to solve malaria or tuberculosis in African populations illuminate or obscure racial inequality? Will biology as a framework continue to trump all else?

At present it appears that whether racialist or postracialist, genomics will amplify its reputation as the apex of all racial science. As society seeks permanent solutions to questions of history, identity, and equality, the government and the general public will continue to enlarge their investment in genomics. In the media, genomic solutions are ever more trumpeted as a deeper form of truth—as in the case of instructional films reflecting on the meaning of race in the wake of the Human Genome Project and televised searches for a person's roots. This is why no matter how much attention scientists put toward solving the "race problem," it persists as a central theme for the field. The paradox is this: the more genomicists attempt to create a science attentive to the social needs around racial equality, the more their promises of a future without race becomes meaningless. It is the widespread desire for a permanent solution to race relations that breeds a biosociality steeped in biological determinism.

Seeking some unending intrinsic value for a taxonomy to be used as a public solution to racial conundrums is shortsighted and problematic. First, it hides the fact that science taxonomies are always under construction, and more importantly, that they are always *politically referential*. While it might remain easy to see that descriptors like "African American" have clear political histories and only make sense in relation to other culturally salient terms such as "white" and "Latino," other descriptors, like "Bantu" or "Yoruba," become fixed conceptual references and dense political transfer points as they become naturalized as being genetic.[47] Second, such biological determinism also hides the fact that race in the general public is always under construction and *politically relational*. Race is a moving target that must be treated in terms of its dynamic formations. As genetic determinism becomes a defining principle for biosociality, social explanations that showcase the messy processes behind racial identity construction and racialization will find less and less of an audience even as these processes continue to unfold.

It would be better then to ask less of genomic classification. We should remain skeptical about how scientific taxonomies can "resolve" long-standing political problems, and scientists for their part should more publicly recognize the limitations of their taxonomies. It would help to highlight the impermanence of any taxonomy and to openly acknowledge the varying political conditions that shape them; we need to remain wary of adhering to commonsense taxonomies—those that make sense within a specific paradigm—since as we have seen, paradigms are ever so fleeting. The proliferation of multiple levels of taxonomy in individual studies—taxonomies for sampling, analyzing data, and reporting findings—and their detailed definition in the science literature would be of most use. When studies are publicized in the wider media, clear explanations of these different levels of classification would limit haphazard racial interpretations in the press. It is the responsibility of editors and reporters as much as authors to make sure such policies are followed.

Such responsibilities are also shared by others, especially experts who work on race from a sociological angle. Often criticism of genomic racial science is waged in the form of a turf war in which social scientists claim to have superior racial optics and measures. Yet privileging a specific racial concept or protocol is what got us where we are today. *All* working models of race—whether those emanating from sociological or biological domains—must be examined. Claiming autonomy for demographic categories in any case is no longer possible in social science, as genomics enters further into social epidemiology and bioethics. In genomic epidemiology, for example, social research that aims to track incidence over time must remain aware of the historical contingency of any categories it uses. This means calibrating categories to new historical contexts and justifying the use of race and specific taxonomies in each and every moment of research.

Social researchers may start asking themselves the same critical questions they often pose to biology: Why is race a variable in my research? Did my subjects self-identify racially because I imposed that categorical framework, or did they speak freely about themselves in racial terms? How do descriptors of self-identity relate to the variables I am interested in? Am I studying race in general or a particular set of categories? Am I transporting categories between different analytical and social contexts? What are the biological, cultural, and sociological assumptions behind the categories I deploy?

Reflection on official demographic variables, such as those mandated by systems of governance, is also needed. As I have shown, the U.S. federal taxonomy has increasingly permeated domestic and international life. In the social sciences and public health, OMB classifications have served as the "go-to" system for a wide array of fields. By virtue of their easy interchangeability with the commonsense continental groupings and their hegemonic status across the sciences and the U.S. public, these classifications stand to be reified across disciplines, social domains, and even racial paradigms. Thus, we must be wary when the state uses the federal taxonomy for registering things apart from racial discrimination. We must also question whether the extension of these categories to new domestic and international research contexts is necessary.

Instead of making a blanket argument against using taxonomies, my research among genomicists has led me to argue that taxonomies need always to be properly defined, and their limits and contexts clearly represented. In this way, an awareness of taxonomies in both research and racial politics as always *contingent* can coexist with the potential for meaningful social change.[48] Our value frameworks will continue to shift with the context of social relations, thus demanding different measures for engendering equality. As Kenan Malik has noted, "the irony is that in order to study human genetic diversity, scientists need socially defined categories of difference."[49] The fact that most human classificatory systems used in science are taken from the social realm needs to be more clearly articulated.

Further, as we open the dialogue to more stakeholders, we need to remain vigilant in checking the hierarchies that shape such discussions. In a biosocial world, where science and politics are more conjoined than ever, ethics are "integral to the conceptualization of research."[50] Yet ethics derived from a society where the value of studying the "bio" and the "socio" is uneven are insufficient. What we need are more forums in which to have "transformative conversations,"[51] where people's lived experiences are represented, and where true deliberation, criticism, and self-examination can occur. Closer collaboration between the representatives of the "bio" and "socio" is needed, but asymmetries need to be addressed before even reaching the discussion table.

Sociological research that situates scientific work within the broader sociopolitical framework is more urgently needed than ever. As I have shown, the structural forces that construe science as outside of sociopolitical contexts are what has led to the reentrenchment of race-positivity in the first place. And these same structural forces have impeded scientists and their critics from seeing how this race-positive science is part and parcel of a particular moment of our history. All too often scientists' actions are taken as evidence of their individual intentions—either racist or antiracist—instead of being interpreted as congruent with a broader sociopolitical dynamic, such as that which has produced at this moment in history a concerted interest in the biology of race.

With a rational unpacking of the political confluences that engender genomic expertise, we might begin envisioning the world differently. No longer will we assume that science exists outside of politics, or that scientific research is the product of individual scientists' own doings. Instead, we will start to make science accountable for the larger trends that shape it. Part of scientists' training is to cultivate an interest in and even quest for critique. And many social activists within the broader community are well positioned to engage with genome scientists and raise consciousness. What we need is a reconceptualization of how scientists are socially situated in order to make such a collaboration productive. At present, stakeholders in genomic research are avid participants in a number of messy and volatile dialogues; many are putting their reputations on the line in order to address the issue of race and to create meaningful social change. Yet many have done so without fully grasping their own embeddedness within a larger sociopolitical movement that is seeking biological answers to social problems. And their arguments have fallen on the ears of others who are not cognizant of the motivations behind this move toward biosociality. With greater attention to its sociological causes and consequences, we can begin to better understand why matters of difference get raised as they do at certain historical junctures. Only then might we begin to produce a congruence of interests toward generating a self-aware and socially beneficial human science.

Notes

Introduction

1. Natalie Angier, "Do Races Differ? Not Really, Genes Show," *New York Times* (Aug. 22, 2000), 1; David Chandler, "Heredity Study Eyes European Origins," *Boston Globe* (May 10, 2001), A22; Sheryl Gay Stolberg, "Shouldn't a Pill Be Colorblind?" *New York Times* (May 13, 2001).

2. As Snait B. Gissis observed, even at the height of genomicist antiracial statements, articles on the genetics of race were multiplying. "When Is 'Race' a Race? 1946–2003," *Studies in History and Philosophy of Biological and Biomedical Sciences* 39 (2008), 437–50. My own Boolean search of [rac* and (gene* or geno*)] in the over three thousand journals in Stanford University's database found a lasting rise in such articles into the twenty-first century.

3. In this book, I home in on genomic work on race. Race is an open referent for systems of meanings that people attribute to physical and social patterns. For many centuries, scientists and philosophers have characterized such patterns in continental terms. For analysis of the connections between genomic concepts of race and ethnicity, see Michael Montoya, *Making the Mexican Diabetic: Race, Science, and the Genetics of Inequality* (Berkeley: University of California Press, 2011). On the broader relationship between race and ethnicity, see Richard Jenkins, *Rethinking Ethnicity: Arguments and Explorations* (Thousand Oaks, CA: Sage Press, 1997).

4. See Audrey Smedley, *Race in North America: Origin and Evolution of a Worldview*, 2nd edition (Boulder, CO: Westview Press, 1999); Nancy Stepan, *The Idea of Race in Science: Great Britain 1800–1960* (London: Macmillan, 1982); Pat Shipman, *The Evolution of Racism: Human Differences and the Use and Abuse of Science* (Cambridge, MA: Harvard University Press, 2002); Stephen Jay Gould, *The Mismeasure of Man* (New York: Norton, 1996); and William H. Tucker, *The Science and Politics of Racial Research* (Chicago: University of Illinois Press, 1996).

5. Examples include physician Josiah Nott's role in Alabama medicine and eugenicist Charles Davenport's leadership of the Immigration Restriction League. Reginald Horsman, *Josiah Nott of Mobile: Southerner, Physician, and Racial Theorist* (Baton Rouge: Louisiana State University Press, 1987); Daniel Kevles, *In the Name of Eugenics* (Berkeley: University of California Press, 1985).

6. Jenny Reardon, *Race to the Finish* (Princeton, NJ: Princeton University Press, 2005).

7. Not surprisingly, this move to colorblind science was a signal of larger political changes. For a discussion of colorblind ideology, see Michael K. Brown, *Whitewashing Race: The Myth of a Colorblind Society* (Berkeley: University of California Press, 2003); Tim Wise, *Colorblind: The Rise of Post-racial Politics and the Retreat from Racial Equity* (San Francisco: City Lights Books, 2010).

8. Nicholas Wade, "For Genome Mappers, the Tricky Terrain of Race Requires Some Careful Navigating," *New York Times* (July 20, 2001); Nicholas Wade, "Race Is Seen as Real Guide to Track Roots of Disease," *New York Times* (June 30, 2002); William Pfaff, "Race Reemerges in a PC World," *Boston Globe* (Aug. 17, 2002), A11.

9. Nicholas Wade, "A Dissenting Voice as the Genome Is Sifted to Fight Disease," *New York Times* (Sept. 15, 2008); "A Revolution at 50, Eric Lander," *New York Times* (Feb. 25, 2003); "Will Self + Spencer Wells," *Seed* (Feb. 4, 2008). Likewise, a series of articles reporting an analysis of double-helix discoverer James D. Watson's genomic makeup made exclamations like: "DNA Pioneer James Watson Is Blacker Than He Thought" (Jonathan Leake, *Sunday Times* [Dec. 9, 2007]) and "Revealed: Scientist Who Sparked Racism Row Has Black Genes" (Robert Verkaik, *Independent* [Dec. 10, 2007]). These headlines flew in the face of the field's overall conjecture that there are no "black" or "white" genes.

10. In the *New York Times*, see David Berreby, "How, but Not Why, the Brain Distinguishes Race," *New York Times* (Sep. 5, 2000); Nicholas D. Kristof, "Is Race Real?" *New York Times* (July 11, 2003); and Brent Staples, "Why Race Isn't as 'Black' and 'White' as We Think," *New York Times* (Oct. 3, 2005). Compare to Peter Kihss, "'Benign Neglect' on Race Is Proposed by Moynihan," *New York Times* (Mar. 1, 1970), 1; and M. William Howard, "Learning to Talk of Race," *New York Times Magazine* (Sept. 6, 1992). Orlando Patterson also lamented this change in "A Poverty of the Mind," *New York Times* (Mar. 26, 2006).

11. Search performed Aug. 11, 2010. Other search terms I used include "racial science," "race science," and "biology of race."

12. The most visible case concerns *New York Times* science writer (former editorial writer and science editor) Nicholas Wade. After covering genomics' early evolution, Wade wrote a best-selling book on human genomics and evolution entitled *Before the Dawn: Recovering the Lost History of Our Ancestors* (New York: Penguin Group, 2006). In his book, he argued that humans have evolved into races because each continent provoked different adaptations. Also see Steve Olson, *Mapping Human History: Discovering the Past Through Our Genes* (Boston: Houghton Mifflin, 2002).

13. Dorothy Nelkin and M. Susan Lindee, *The DNA Mystique: The Gene as a Cultural Icon* (Ann Arbor: University of Michigan Press, 2004).

14. In the context of race, biological essentialism is the idea that each race possesses unique, immutable, mutually exclusive, and heritable biological qualities.

15. In *Inclusion: The Politics of Difference in Medical Research* (Chicago: University of Chicago Press, 2007), Steven Epstein shows how New Left and feminist advocacy groups successfully petitioned the federal government to create inclusionary policies to replace a colorblind, one-size-fits-all standardized research model. He has called this transformation a "biopolitical paradigm shift" to draw attention to the confluence of state and scientific interests.

16. Jonathan Kahn, "Genes, Race, and Population: Avoiding a Collision of Categories," *American Journal of Public Health* 96 (2006), 1965–70; Dorothy Roberts, "Is Race-Based Medicine Good for Us? African American Approaches to Race, Biomedicine, and Equality," *Journal of Law, Medicine and Ethics* 36 (2008), 537–45; *Fatal Invention: How Scientists, Politics, and Big Business Re-create Race in the Twenty-First Century* (New York: New Press, 2011); Sandra Soo-Jin Lee, "The Ethical Implications of Stratifying by Race in Pharmacogenomics," *Clinical Pharmacological Therapeutics* 81 (2007), 122–25.

17. In this book, I refer to the dominant conception of race as the "lay conception" or "common lay notion" of race. This dominant conception posits a black-white binary, and in America includes ideas about Native American, Asian, and Latino biological and cultural uniqueness. An example of a continent-based system of ancestry is: African, Native American, Asian, European, and Pacific Islander. Two examples of common racial terms scientists reported working with are "African American" and "Caucasian."

Nonbiological population concepts and fluid or hybrid notions of race also circulate in the broader society. For the purposes of my analysis, I focus on the above, most prevalent public perception of race. On lay conceptions of race and genetics, see the work of Celeste Michelle Condit.

18. David Skinner and Paul Rosen, "Opening the White Box: The Politics of Racialised Science and Technology," *Science as Culture* 10 (2001), 285–300; Priscilla Wald, "Race Based Genomics Research: Not So Black and White," *Ethos* (2004), 12–14; Alexandra E. Shields, Michael Fortun, Evelynn M. Hammonds, Patricia A. King, et al., "The Use of Race Variables in Genetic Studies of Complex Traits and the Goal of Reducing Health Disparities: A Transdisciplinary Perspective," *American Psychologist* 77 (2005), 77–103.

19. Richard Tutton, "Opening the White Box: Exploring the Study of Whiteness in Contemporary Genetics Research," *Ethnic and Racial Studies* 30 (2007), 557–69; Duana Fullwiley, "The Biologistical Construction of Race: 'Admixture' Technology and the New Genetic Medicine," *Social Studies of Science* 38 (2008), 695–735; "Race and Genetics: Attempts to Define the Relationship," *BioSocieties* 2 (2007), 221–37; "The Molecularization of Race: Institutionalizing Human Difference in Pharmacogenetics Practice," *Science as Culture* 16 (2007), 1–30; Linda M. Hunt and Mary S. Megyesi, "The Ambiguous Meanings of the Racial/Ethnic Categories Routinely Used in Human Genetics Research," *Social Science and Medicine* 66 (2008), 349–61.

20. In Chapter Three, I discuss science activism in the context of a new biologically focused social order. Another way to think about scientists' performance of civic duties is to identify their role in engaging certain civic epistemologies. Sheila Jasanoff,

Designs on Nature: Science and Democracy in Europe and the United States (Princeton, NJ: Princeton University Press, 2007).

21. Fullwiley, "Biologistical Construction of Race"; Alondra Nelson, "Bio Science: Genetic Genealogy Testing and the Pursuit of African Ancestry," *Social Studies of Science* 38 (2008), 759–83; "The Factness of Diaspora," in *Revisiting Race in a Genomic Age*, edited by Barbara Koenig, Sandra S.J. Lee, and Sarah S. Richardson (Piscataway, NJ: Rutgers University Press, 2008).

22. Sharon Traweek, *Beamtimes and Lifetimes* (Cambridge, MA: MIT Press, 1988); Evelyn Fox Keller, *Reflections on Gender and Science* (New Haven, CT: Yale University Press, 1984). See Banu Subramanian's treatment of the problem this poses for feminist science and engineering mentors in "Assimilating the 'Culture of No Culture' in Science: Feminist Interventions in (De)Mentoring Graduate Women," *Feminist Teacher* 12 (1998), 12–28.

23. Paul Rabinow has written extensively on ethos in science. For a theoretical treatment of the topic, see *Anthropos Today* (Princeton, NJ: Princeton University Press, 1996). For an ethnographic analysis, see *A Machine to Make a Future* (Princeton, NJ: Princeton University Press, 2003).

24. The reverse is also true: interviewees interpret their pasts with concerns of the present. On the relationship between past and present knowledge, see Arthur Coleman Danto, *Narration and Knowledge: Including the Integral Text of Analytical Philosophy of History* (New York: Columbia University Press, 1985).

25. Reardon locates the origins of such a stance in the postwar period, when contributors to the UNESCO Statements on Race attempted to create an apolitical definition of race as genetic subspecies. Although they stressed the importance of cultural variation and mutability, these postwar scientists actually produced a more biologically deterministic definition of race as population or subspecies difference than the eugenics-based scientific racism that preceded it. When arguing that perceived racial differences were actually caused by ethnic mores and intergenerational social hierarchies, these geneticists charged others to separate the term "race" from anything *but* fixed inherited traits. Similarly, during the IQ debates of the 1970s, as prominent geneticists passionately argued against the notion of inbred intellectual aptitude, they continued to refine the biological definition of race as subspecies. These scientists established a range of continental "-oid" terms that have since become deeply ingrained in the academic and public lexicons. Their most prominent articles of the time focused on the question of "racial divergence," or the evolutionary split between continental populations. Reardon argues that the Diversity Project was devised as an application of the latest genomic technologies to this age-old question (Reardon, *Race to the Finish*).

26. For uses of the phrase "the decade of the genome," see Clare Garvey, "Ten Years of Genome Biology," *Genome Biology* 11 (2010), 101; and Francis Collins, "Has the Revolution Arrived?" *Nature* 464 (2010), 674–75.

27. In recent years, social scientists have produced the first interviews-based ethnographies of race and genomics. Amade M'Charek studied population construction in two Human Genome Diversity Project labs (Amade M'Charek, *The Human Genome*

Diversity Project: An Ethnography of Scientific Practice [Cambridge, UK: Cambridge University Press, 2005]). Reardon explored the sociological reasons for the rise and fall of a human genome project dedicated to mapping human variation (Reardon, *Race to the Finish*). Michael Montoya and Duana Fulwilley performed lab studies of government-sponsored health research centers in order to get an insider's view on everyday classificatory practice (Montoya, *Making the Mexican Diabetic;* Fullwiley, "Biological Construction of Race"; "Race and Genetics"; "Molecularization of Race"). Taking a multisited transcultural approach, Ian Whitmarsh and Kim Tallbear examined the production of genomic knowledge about race in a global context (Ian Whitmarsh, *Biomedical Ambiguity: Race, Asthma, and the Contested Meaning of Genetic Research in the Caribbean* [Ithaca, NY: Cornell University Press, 2008]; Kimberly Tallbear, "Native American DNA: Narratives of Origin and Race," Ph.D. dissertation [Department of History of Consciousness, University of California, Santa Cruz, 2005]). Like all of these studies, my research is simultaneously local and global in scope. Though all of the scientists I studied conduct multinational research, the majority have been educated in the United States or are based in North America. American policies and markets most closely shape their research, even for those based outside the United States.

28. See Harry M. Collins, *Changing Order: Replication and Induction in Scientific Practice*, 2nd edition (Chicago: University of Chicago Press, 1992); Andrew Pickering, *Constructing Quarks: A Sociological History of Particle Physics* (Chicago: University of Chicago Press, 1984); and Steven Shapin and Simon Schaffer, *Leviathan and the Air-Pump: Hobbes, Boyle and the Experimental Life* (Princeton, NJ: Princeton University Press, 1985).

29. Prior to my ethnographic work, I performed content analysis of five venues where articles related to the genomics of race appear: *American Journal of Human Genetics, Journal of the American Medical Association, Nature, Nature Genetics,* and *New England Journal of Medicine*. From these journals I gleaned key publications of the field, such as the draft map of the human genome and Phase I of the International HapMap Project, the introduction of field-defining bioinformatic programs and databases, and human genomics policy commentaries. I created my sample from lead and senior authors of these articles. I invited scientists to be interviewed with a brief e-mail denoting my interest in discussing race and science in the wake of the Human Genome Project.

30. Bruno Latour, *Science in Action: How to Follow Scientists and Engineers Through Society* (Cambridge, MA: Harvard University Press, 1987).

31. My archive consisted of the 732 articles constitutive of genomic literature and race policy published between 1986 and 2010. Using Stanford University's SULAIR/Lane Medical Library, New York University, and Brown University electronic databases, I performed individual Boolean searches on rac* and gene* or geno* (over three thousand scientific journals). For the journals that did not offer a Boolean term* algorithm, I typed in "race", "racial", "gene", "genetic", "genome", and "genomic" in the *anywhere in the text* search mode. I cross-performed this search through two databases: PubMed@ Stanford and BIOSIS. I searched on "gene", "genetic", "genome", and "genomic" in

Science, Nature, Journal of the American Medical Association, and *New England Journal of Medicine*; archived all articles pertaining to human variation; and used their reference lists to reach other leading articles on human variation. In addition to scientific articles, I collected all 2000–2010 magazine articles and news transcripts mentioning "race" or "racial" and "gene" or "genetic" or "genome" or "genomic" using Factiva. I cross-searched magazine and television sources on their respective websites (ABC.com, CBS.com, CNN.com, etc.). I used Google Scholar (http://scholar.google.com) to collect all articles written by the scientists interviewed here. Finally, at Stanford University's Jackson Library, I compiled biotechnology and pharmaceuticals industry reports and collected the annual reports of an array of genetic genealogy and personal genomics companies.

32. David O. Edge and Michael J. Mulkay, *Astronomy Transformed: The Emergence of Radio Astronomy in Britain* (New York: Wiley, 1976); Ian I. Mitroff, *The Subjective Side of Science* (Amsterdam: Elsevier Scientific, 1974); "Norms and Counter-norms in a Select Group of Apollo Moon Scientists: A Case Study of the Ambivalence of Scientists," *American Sociological Review* 395 (1974), 79–95; Michael J. Mulkay, "The Sociology of Science in the West," *Current Sociology* 28 (1980), 1–184; *Science and the Sociology of Knowledge* (London: Allen & Unwin, 1979); "Norms and Ideology in Science," *Social Science Information* 15 (1976), 637–56. Also see the work of Derek J. de Solla Price.

33. Paul Rabinow, *French DNA* (Chicago: University of Chicago Press, 1999), 17–18.

34. Steven Shapin, *The Scientific Life: A Moral History of a Late Modern Vocation* (Chicago: University of Chicago Press, 2008), 312.

35. Rabinow, *Essays on the Anthropology of Reason,* 244. Also see *French DNA,* 13.

36. Nikolas Rose, *The Politics of Life Itself* (Princeton, NJ: Princeton University Press, 2007).

37. *Ibid.,* 22–23; and Rabinow, *French DNA,* 16.

38. *Ibid.,* 26–27.

39. Reardon, *Race to the Finish*; Montoya, *Making the Mexican Diabetic*; Tutton, "Opening the White Box."

40. Critics of colorblindness believe that ignoring race does nothing to address institutional racism, such as the uneven transfer of wealth and privilege between generations.

41. Fullwiley, "Biologistical Construction of Race"; Nelson, "Bio Science"; "Factness of Diaspora."

42. Reardon's interviews with staff scientists at three personal genomics companies have depicted genomicists as nimbly ascribing new characteristics to their consumer base—such as building them as educated and genomics-savvy—in the race to produce a clear consumer target. Yet this analysis does not focus on the scientists' subjectivity as much as the way early user identities are coproduced by company scientists and users themselves in the shared interest of bringing new products to market. Jenny Reardon, "The 'Persons' and 'Genomics' of Personal Genomics," *Personalized Medicine* 8 (2011), 95–107.

43. Michael A. Fortun, *Promising Genomics: Iceland and deCODE Genetics in a*

World of Speculation (Berkeley: University of California Press, 2008); Kaushik Sunder Rajan, *Biocapital: The Constitution of Postgenomic Life* (Durham, NC: Duke Univerity Press, 2006); Daniel Lee Kleinman, *Impure Cultures: University Biology and the World of Commerce* (Madison: University of Wisconsin Press, 2003); Sarah Franklin, *Dolly Mixtures* (Durham, NC: Duke University Press, 2007). I discuss financial incentives for genomicists to use a racial taxonomy in the following chapter.

44. Roberts, *Fatal Invention*, 149.

45. Shapin, *Scientific Life*.

46. Paul Rabinow and Gaymon Bennett, "Human Practices: Interfacing Three Modes of Collaboration," in *The Ethics of Protocells: Moral and Social Implications of Creating Life in the Laboratory*, edited by Mark Bedau and Emily C. Parke (Cambridge, MA: MIT Press, 2009).

47. Carlos Novas, "The Political Economy of Hope: Patients' Organizations, Science and Biovalue," *BioSocieties* 1 (2006), 289–305; Sahra Gibbon and Carlos Novas, *Genetics, Biosociality and the Social Sciences: Making Biologies and Identities* (London: Routledge, 2007); Carlos Novas and Nikolas Rose, "Genetic Risk and the Birth of the Somatic Individual," *Economy and Society* 29 (2000), 495–513.

48. William Edward Burghardt Du Bois, *The Souls of Black Folk* (New York: Penguin, 1995 [1903]).

49. Paul Gilroy, *The Black Atlantic: Modernity and Double Consciousness* (Cambridge, MA: Harvard University Press, 1992); Amade M'Charek, "Fragile Differences, Relational Effects: Stories About the Materiality of Race and Sex," *European Journal of Women's Studies* 17 (2010), 307–22.

50. Ian Hacking, "Making Up People," in *The Science Studies Reader*, edited by Mario Biagioli (New York: Routledge, 1999), 161–71.

51. See, for example, Tony N. Brown and Chase L. Lesane Brown, "Race Socialization Messages Across Historical Time," *Social Psychology Quarterly* 2 (2006) 201–13; and Michael Omi and Howard Winant, *Racial Formations in the United States* (New York: Routledge, 1994).

52. Epstein, *Inclusion*.

53. George Lipsitz, *The Possessive Investment in Whiteness: How White People Profit from Identity Politics* (Philadelphia: Temple University Press, 2006).

54. *Ibid.*, vii.

55. Nadia Abu El-Haj, "The Genetic Reinscription of Race," *Annual Review of Anthropology* 36 (2007), 283–s300.

56. On the other hand, knee-jerk acceptance of genomic findings also holds thin. It has become fashionable to use genomic arguments to bolster the belief that race is not "in the genes" and that social equality is the true state of nature. However, I will show that even those claims most lauded by a society that hungers for social justice are themselves conditioned by malleable circuits of assumptions and norms.

57. Kwame Anthony Appiah, "*Racisms*," in *Anatomy of Racism*, edited by David Theo Goldberg (Minneapolis: University of Minnesota Press, 1990), 222; George M. Frederickson, *Racism: A Short History* (Princeton, NJ: Princeton University Press, 2002),

153–54; Paul Gilroy, *Against Race: Imagining Political Culture Beyond the Color Line* (Cambridge, MA: Belknap Press of Harvard University Press, 2000).

58. Nancy Fraser, *Redistribution or Recognition? A Political-Philosophical Exchange* (New York and London: Verso Books, 2003); Charles Taylor, "The Politics of Recognition," in *Race and Ethnicity: Comparative and Theoretical Approaches*, edited by John Stone (Hoboken, NJ: Wiley-Blackwell, 2003), 377–81.

59. Genetics has arguably always been an activist field. For analysis of the relationship between racial genetics and eugenic policies, see Kevles, *In the Name of Eugenics*; and Tucker, *The Science and Politics of Racial Research*. I limit this introduction to the antiracist activism that emerged in the postwar period.

60. As historians and philosophers of genetics have argued, this populational definition of race was no less racial than previous racist ones. In fact, it buttressed the notion of essential biological races even as it denied a hierarchy to those races. Jenny Reardon, "Decoding Race and Human Difference in a Genomic Age," *differences* 15 (2004), 3–38; Lisa Gannett, "Racism and Human Genome Diversity Research: The Ethical Limits of 'Population Thinking,'" *Philosophy of Science* 68 (2001), S479–S492. Also see Gissis, "When Is 'Race' a Race?"; Jonathan Marks, *What It Means to Be 98% Chimpanzee* (Berkeley: University of California Press, 2003).

61. Jensen argued that intelligence was 70 percent attributable to differences between races and only 30 percent environmental.

62. Walter F. Bodmer and Luigi L. Cavalli-Sforza, *Genetics, Evolution, and Man* (San Francisco: Freeman, 1976), 563.

63. L. Zonta Sgaramella and Luigi L. Cavalli-Sforza, "Origin of Caucasians Tested Using Genetic-Markers," *Clinical Genetics* 10 (1976), 374; A. Pizza, P. Menozzi, and L. Cavalli-Sforza, "The Making and Testing of Geographic Gene-Frequency Maps," *Biometrics* 37 (1981), 635–59; Masatoshi Nei and A. K. Roychoudhry, "The Genetic Relationship and Evolution of Human Races," *Evolutionary Biology* 14 (1982), 1–59; Masatoshi Nei, "Human Evolution at the Molecular Level," *Population Genetics and Molecular Evolution* (1985), 41–64; Satoshi Horai, Takashi Gojobori, and Ei Matsunaga, "Distinct Clustering of Mitochondrial DNA Types Among Japanese, Caucasians and Negroes," *Japanese Journal of Genetics* 61 (1986), 271–75.

64. Gayatri Chakravorty Spivak, Donna Landry, and Gerald M. McLean, *The Spivak Reader: Selected Works of Gayatri Chakravorty Spivak* (New York: Routledge, 1996); Lisa Lowe, "Heterogeneity, Hybridity, Multiplicity: Marking Asian American Differences," in *A Companion to Asian American Studies* (Oxford, UK: Blackwell, 2005).

65. Sheila Jasanoff calls this transdomain cooperative phenomenon "co-production." See *States of Knowledge: The Co-Production of Science and the Social Order* (New York: Routledge, 2006). It is also worth noting the interdisciplinarity of genomics within the natural sciences, across fields like evolutionary biology, computer science, chemistry, physiology, and more.

66. Fullwiley, "Biologistical Construction of Race"; Montoya, *Making the Mexican Diabetic*.

67. Rose, *Politics of Life Itself*.

Chapter 1

1. *Race, the Floating Signifier* (video), Stuart Hall, host (1996).

2. In her work on the formation of "epistemic habits" among colonial elites, Ann Stoler argues the importance of examining the history of racial governance in order to know how contemporary taxonomies are changing. Ann Laura Stoler, *Along the Archival Grain: Epistemic Anxieties and Colonial Common Sense* (Princeton, NJ: Princeton University Press, 2009), 39.

3. In the interest of brevity, I refer to Directive No. 15's taxonomy as "Census race."

4. Office of Management and Budget (OMB), "OMB Directive No. 15," in *Race and Ethnic Standards for Federal Statistics and Administrative Reporting: Statistical Policy Handbook* (Washington, DC: OMB Publications Office, 1978).

5. Melissa Nobles, *Shades of Citizenship* (Stanford, CA: Stanford University Press, 2000), 80.

6. Since then, Directive No. 15 classifications have been reworked with greater biocontinental explication. For example, in the past two national censuses, "Asian" was defined as "a person with origins in any of the original peoples of the Far East, Southeast Asia, or the Indian Subcontinent." "White" was similarly defined as "a person with origins in any of the original peoples of Europe, the Middle East, or North Africa." "Native Hawaiian or Other Pacific Islander" was also defined in terms of original peoples. Yet "Black or African American" was defined as "a person having origins in any of the black racial groups of Africa." "American Indian or Alaska Native" required that the person "maintains tribal affiliation or community attachment." See Dvora Yanow, *Constructing "Race" and "Ethnicity" in America: Category-Making in Public Policy and Administration* (New York: Sharpe, 2002). Although the last census allowed respondents to mark more than one box—attesting to a potentially plural, noncontinental, operational definition of race—Directive No. 15's pattern of shuttling between geographical ("original peoples"), biological ("racial groups"), and social ("community attachment") definitions indicates the fuzzy and contradictory federal characterization of race. The Government Accountability Office in partnership with the U.S. Census Bureau continues to examine the appropriateness of Directive No. 15 classifications in its census field tests.

7. Although the OMB permits the marking of more than one box, multiple-box respondents are "allocated" to single race categories. Here are the allocation guidelines put forth by the OMB:

> Allocation Guidance: Federal agencies will use the following rules to allocate multiple race responses for use in civil rights monitoring and enforcement.
>
> Responses in the five single race categories are not allocated.
>
> Responses that combine one minority race and white are allocated to the minority race.
>
> Responses that include two or more minority races are allocated as follows: If the enforcement action is in response to a complaint, allocate to the race that the complainant alleges the discrimination was based on. If the enforcement action requires assessing disparate impact or discriminatory patterns, analyze the patterns based on alternative allocations to each of the minority groups. (OMB, 2000)

The Census, conversely, tabulates data in terms of "race *alone*" versus "*two or more races*" populations. Included in the "race *alone*" category are those who marked "Some other race." Thus, OMB and all federal agencies recognize five races, while the Census recognizes seven "populations" (U.S. Census Bureau, 2000) and sixty-three combinations of race:

6 race alone categories

15 categories of 2 races (e.g., White and African American; White and Asian; etc.)

20 categories of 3 races

15 categories of 4 races

6 categories of 5 races

1 category of 6 races

= 63 possible combinations

8. Yanow, *Constructing "Race" and "Ethnicity" in America*, 116–17.

9. Despite the tenuous nature of federal categories, the government's belief that the U.S. citizenry is, and always has been, divisible into racial subgroups holds fast. For example, the Census Bureau has read contemporary racial classification back into time by enumerating the 1790 population by 1990 categories (http://www.census.gov/prod/www/abs/decennial/1790.htm).

10. The next chapter explores reasons for this delay in public health implementation.

11. Epstein, *Inclusion*.

12. U.S. Department of Health and Human Services, *Healthy People 2000: Our Nation's Prevention Agenda*, http://www.crisny.org/health/us/health7.html.

13. *Healthy People 2000: Citizens Chart the Course*, edited by Michael A. Stoto, Ruth Behrens, and Connie Rosemont (Washington, DC: National Academy Press, 1990), 51.

14. National Institutes of Health (NIH), *The NIH Revitalization Act of 1993*, PL 103-43, http://grants.nih.gov/grants/guide/notice-files/not94-100.html. In 1993, CDC also drafted the "Use of Race and Ethnicity in Public Health Surveillance" statement. Centers for Disease Control and Prevention (CDC), "Use of Race and Ethnicity in Public Health Surveillance Summary of the CDC/ATSDR Workshop," *MMWR* 42[RR-10], (June 25, 1993). For a report on CDC implementation of OMB categories, see CDC, *Federal Register* 64, no. 46 (Mar. 10, 1999), 11915–20.

15. Miriam Kelty, Angela Bates, and Vivian W. Penn, "NIH Policy on the Inclusion of Women and Minorities as Subjects in Clinical Research," in *Principles of Practice and Clinical Research*, 2nd edition, edited by John I. Gallin and Frederick P. Ognibene (Burlington, MA: Elsevier, 2002).

16. Food and Drug Administration (FDA), *Investigational New Drug Applications and New Drug Applications*, http://www.fda.gov/oashi/patrep/demo.html.

17. *Ibid.*

18. U.S. Department of Health and Human Services (HHS), *Improving the Collection and Use of Racial and Ethnic Data in HHS*, http://aspe.hhs.gov/datacncl/RaceRpt/execsumm.htm.

19. At this time, *Healthy People 2000* also formulated a minimum template for population-based objectives. It enforced the use of Census race in all Healthy People

programs. Department of Health and Human Services (HHS), *Healthy People 2010 Objectives: Draft for Public Comment* (Washington, DC: HHS, Office of Public Health and Science, 1998).

20. Phase III Clinical Trials are studies wherein a treatment is given to large groups of people, often at multiple sites, to assess efficacy, side effects, usefulness in comparison with other drugs and treatments, and other information. Phase III is the last trial stage before a drug is released on the market.

21. NIH, "Guidelines on the Inclusion of Women and Minorities as Subjects in Clinical Research," http://grants2.nih.gov/grants/guide/notice-files/NOT-OD-00-048.html.

22. NIH, "Policy on Reporting Race and Ethnicity Data: Subjects in Clinical Research," http://grants.nih.gov/grants/guide/notice-files/NOT-OD-01-053.html.

23. "FDA Issues Guidance for Industry on the Collection of Race and Ethnicity Data in Clinical Trials for FDA Regulated Products," Washington, DC: T03-07 (Jan. 23, 2003).

24. FDA, "Guidance for Industry on the Collection of Race and Ethnicity Data in Clinical Trials for FDA Regulated Products," http://www.fda.gov/cder/guidance/index.htm.

25. Ian Whitmarsh has argued that trials on foreign populations have contributed to the U.S. government's concern about data transferability across populations. His study of Barbadian efforts to portray its population as a desirable asthma trial group shows how developing nations vie for a place in the U.S.-dominated pharmaceutical market. A prerequisite for success is to establish a biological argument for continuity between domestic and foreign populations. In this case, the Barbadian government constructed a "black" identity for its populace that could be linked to the "black race" in the United States—a group disproportionately affected by asthma and thus a potential market for race-based medicine (Whitmarsh, *Biomedical Ambiguity*).

26. FDA, "Guidance for Industry . . ."

27. The latter studies are called "family-based linkage analysis" and "candidate gene analysis." Mapping a gene's position on a chromosome is called "positional cloning."

28. See http://www.genome.gov/pfv.cfm?pageID=19518660 for government-funded research projects associated with genome-wide association studies, and http://www.genome.gov/26525384 for a catalog of published genome-wide association studies.

29. Recombination is the process by which DNA is broken and fused with a different piece of DNA. In humans, recombination occurs prior to sex-cell division (meiosis) and involves the mixing of hereditary material from parents. Nonrecombinant portions of the genome (Y-chromosomal, mitochondrial) provide ancestral information, as they are passed down from generation to generation without change.

30. A haplotype is the sequence of a single chromosome, summarized as a unique combination of known polymorphic sites. A haplogroup is a group of similar haplotypes that share a common ancestor with a single nucleotide polymorphism mutation. Haplotypes and haplogroups are represented by a sequence of letters and numbers.

31. Shriver's National Institute of Justice grant was entitled "Isolation of Population-Specific Alleles for Ethnic Affiliation Estimation" (95-IJ-CX-008).

32. Mark Shriver, Michael W. Smith, Li Jin, Amy Marcini, et al., "Ethnic-Affiliation Estimation by Use of Population-Specific DNA Markers," *American Journal of Human Genetics* 60 (1997), 962.

33. As Pilar Ossorio has argued, just because populations are considered to have different genetic frequencies doesn't mean an individual's genotype can be assumed to be consistent with a population's average profile at that locus. Ossorio, "About Face: Forensic Genetic Testing for Race and Visible Traits," *Journal of Law, Medicine, and Ethics* 38 (2006), 278–92.

34. According to Kimberly Tallbear and Deborah Bolnick, some tribes have considered using these and other companies' mitochondrial DNA and Y-chromosome services. Kimberly Tallbear and Deborah Bolnick, "'Native American DNA' Tests: What Are the Risks to Tribes?" *The Native Voice* (Dec. 2004), 3–17. One may raise the point that rights, resources, and representation are actually being offered on a tribal, therefore subcontinental/nonracial, level; yet tests usually only determine whether a person has "Native American," "Asian," "African," or "Indo-European" admixture, to use the terms of one company's example (http: //www.genelex.com). Since many companies do not test for tribal or subcontinental variation, the information they sell is better described as racial membership.

35. Amy Harmon, "Seeking Ancestry in DNA Ties Uncovered by Tests," *New York Times* (Apr. 12, 2006).

36. Joan Fujimura and Ramya Rajagopalan, "Different Differences: The Use of 'Genetic Ancestry' Versus Race in Biomedical Human Genetic Research," *Social Studies of Science* (Dec. 7, 2010), http://sss.sagepub.com/content/early/2010/11/30/0306312710379170.

37. On genome projects in the developing world, see Ruha Benjamin, "A Lab of Their Own: Genomic Sovereignty as Postcolonial Science Policy," *Policy and Society* 28 (2009), 341–55. On biovalue, see Catherine Waldby, "Stem Cells, Tissue Cultures and the Production of Biovalue," *Health: An Interdisciplinary Journal for the Social Study of Health, Illness and Medicine* 6 (2002), 305–23; Catherine Waldby and Robert Mitchell, *Tissue Economies: Blood, Organs, and Cell Lines in Late Capitalism* (Durham, NC: Duke University Press, 2006); Charis Thompson, *Making Parents: The Ontological Choreography of Reproductive Technology* (Cambridge, MA: MIT Press, 2005); Melinda Cooper, *Life as Surplus: Biotechnology, and Captialism in the Neoliberal Era* (Seattle: University of Washington Press, 2008); and Herbert Gottweiss, Brian Salter, and Catherine Waldby, *The Global Politics of Human Embryonic Stem Cell Research* (London: Palgrave, 2008).

38. Cancer Genome Project, http://www.sanger.ac.uk/genetics/CGP; Cancer Genome Atlas, http://cancergenome.nih.gov.

39. An assessment of 2008 fiscal trends suggested that this meta-agency was the only thriving part of the NIH, amidst the budget setbacks of the Bush administration; see Jocelyn Kaiser, "NIH Hopes for More Mileage from Roadmap," *Science* 319 (2008), 716. Since its founding, genomics has been consistently at the forefront of all twenty-seven institutes' strategic plans.

40. Andrew Pollack, "Awaiting the Big Payoff," *New York Times* (June 15, 2010), B1.

41. *Ibid.*

42. Jonathan Kahn, Mandating Race: How the USPTO Is Forcing Race into Biotech Patents," *Nature Biotechnology* 29 (2011), 401–3.

43. Jennifer Reineke Polhaus and Robert M. Cook-Deegan, "Genomics Research: World Survey of Public Funding," *BMC Genomics* 9 (2008), 472.

44. Even the latest Phase III collection concentrates on populations with these origins. See http://www.sanger.ac.uk/resources/downloads/human/hapmap3.html.

45. Adriana Petryna, *When Experiments Travel: Clinical Trials and the Global Search for Human Subjects* (Princeton, NJ: Princeton University Press, 2009), 10–46.

46. Whitmarsh, *Biomedical Ambiguity.*

47. For analysis of the U.K. national biobank and population genomics in the United Kingdom, see Richard Tutton, "Biobanks and the Inclusion of Racial/Ethnic Minorities," *Race/Ethnicity: Multidisciplinary Global Perspectives* 3 (2009), 75–95; and Andrew Smart, Richard Tutton, Richard Ashcroft, Paul A. Martin, et al., "Can Science Alone Improve the Measurement and Communication of Race and Ethnicity in Genetic Research? Exploring the Strategies Proposed by Nature Genetics," *Biosocieties* 1 (2006), 313–24. For a review of population-biobank literature, see Oonagh Corrigan and Richard Tutton, "Biobanks and the Challenges of Governance, Legitimacy and Benefit," in *The Handbook of Genetics and Society: Mapping the New Genomic Era*, edited by Paul Atkinson, Peter E. Glasner, and Margaret M. Lock (New York: Routledge, 2009).

48. Pollack, "Awaiting the Big Payoff."

49. Emily Singer, "Sequencing Price Drops Even Lower: Start-up Complete Genomics Reveals Three New Human Genomes," *Technology Review* (Nov. 6, 2009), http://www.technologyreview.com/biomedicine/23891/?a=f.

Still, many communities around the world continue to live without basic health needs being met, let alone having access to genomic technologies. While some have called on genomic knowledge and its ubiquitous availability as a solution—marking genomics as a human right—others warn of the inherent biases in matters of genomic science. Pointing to genomics' narrow focus on diseases that affect but 10 percent of the world's population, critics advocate for alternative public health enquiry. World Health Organization (WHO), *Genomics and the Global Health Divide*, http://www.who.int/genomics/healthdivide/en/index.html.

50. Academic laboratories and privately funded laboratories are also conducting large-scale sequencing projects. Harvard's George Church heads up the Personal Genome Project, a private project dedicated to publishing the sequence of ten individual genomes. Craig Venter is preparing to publish the diploid sequence of ten individuals and plans to sequence ten thousand diploid genomes in the next ten years. Elizabeth Pennisi, "Number of Sequenced Human Genomes Doubles," *Science* 322 (2008), 838.

51. The Beijing Genome Institute plans to sequence ninety-nine more genomes to better understand Chinese lineages. *Ibid.*; Jun Wang, Wei Wang, Ruiqiang Li, Yingrui Li, et al., "The Diploid Genome Sequence of an Asian Individual," *Nature* 456 (2008), 60–66.

52. Martin Enserink, "Read All About It—The First Female Genome! Or Is It?" *Science* 320 (2008), 1274.

53. Pennisi, "Number of Sequenced Human Genomes Doubles." Curiously, in the same issue, just below this article, reporter Constance Holden reported on the Genome Research Institute's unanimous agreement that terms like "Caucasian" are completely unacceptable for science reporting. This shows the relentlessness of the continental framework in contemporary genomics, despite warnings.

54. The 1000 Genomes Consortium explains its decision to use HapMap samples in terms of consent issues:

> The Samples/ELSI Group focused initially on issues of informed consent and the nature (number and diversity) of the DNA samples required for the Project. After much work, this Working Group has defined standards for informed consent for samples to be incorporated into the Project, approved the use of the already collected extended set of HapMap samples, and is now focused on identifying additional samples that need to be obtained for the Project. Aravinda Chakravarti, co-chair of the Samples/ELSI Group, has been working with the Analysis Group to understand the population genetic parameters that will guide design of the overall sample set for the full Project. (1000 genomes consortium, council letter [Sept. 4, 2008], http://www.1000genomes.org/page.php?page=about)
>
> Of particular importance to the project are the syncing of 1000 genomes protocols with the available HapMap data and the population frameworks of 1000 Genomes' three pilot projects. (1000 genomes consortium, "Meeting Report: A Workshop to Plan a Deep Catalog of Human Genetic Variation, Executive Summary" [Cambridge, UK, Sept. 17–18, 2008], http://www.1000genomes.org/page.php?page=about)

55. *Ibid.* The 1000 Genomes Project will surpass the International HapMap Project's survey of the common markers of human variation present in at least 5 percent of the population to "achieve a nearly complete catalog of common human genetic variants (defined as frequency 1% or higher) by generating high-quality sequence data for >85% of the genome for three sets of 400–500 individuals." It has turned its funding efforts toward a complex of research programs at four sequencing centers across the United States. These centers, used for the International HapMap Project, are: Baylor College of Medicine Human Genome Sequencing Center (www.hgsc.bcm .tmc.edu), Broad Institute Genome Sequencing Center (www.broad.mit.edu), Washington University, St. Louis Genome Sequencing Center (www.genome.wustl.edu), and National Institutes of Health Intramural Sequencing Center (www.nisc.nih.gov). The total amount of funds for 2006 registered at $130 million. National Human Genome Research Institute (NHGRI), "The Large-Scale Genome Sequencing Program" (2008), http://www.genome.gov/pfv.cfm?pageID=10001691. Well over two hundred sequencing targets have been approved in a framework of five main research programs (Human Medical Sequencing, Human Genetic Variation, Comparative Sequencing to Understand the Human Genome, Sequencing the Human Microbiome, and Pathogens and Vectors).

56. This actually contradicts the foundational view that Africans are more diverse than any other continental population in the world.

57. Jonathan Kahn, "Beyond BiDil: The Expanding Embrace of Race in Biomedical Research and Product Development," *Saint Louis University Journal of Health, Law,*

and Public Policy 3 (2010), 69. Critics have also questioned the exploitative effects of a severely more expensive combination drug. BiDil's only benefit beyond taking the generic drugs in combination was its reduction of the pill burden. BiDil would thus lead to African Americans paying more for the same medication.

58. *Ibid.*

59. *Ibid.* Also see "AutoGenomics CYP 2C19 Test Used in Study to Personalize Clopidogrel (Plavix) Dosing in Patients," *Medical News Today* (Jan. 23, 2009).

60. Epstein has referred to this problem in terms of "niche standardization," or a way of standardizing humans at the level of the social group for scientific scrutiny, political administration, or marketing. Epstein, *Inclusion*, 135.

61. http://plato.stanford.edu/entries/identity-politics.

62. Steven Epstein, *Impure Science: AIDS, Activism, and the Politics of Knowledge* (Berkeley: University of California Press, 1998); Alondra Nelson, *Body and Soul: The Black Panther Party and the Fight Against Medical Discrimination* (Minneapolis: University of Minnesota Press, 2011).

63. M. Susan Lindee, *Moments of Truth in Genetic Medicine* (Baltimore: Johns Hopkins University Press, 2008).

64. Keith Wailoo and Stephen Pemberton, *The Troubled Dream of Genetic Medicine: Ethnicity and Innovation in Tay-Sachs, Cystic Fibrosis, Sickle Cell Disease* (Baltimore: Johns Hopkins University Press, 2006).

65. Henry Louis Gates, *Finding Oprah's Roots, Finding Your Own* (New York: Random House, 2007).

66. His recent television project, *Faces of America*, has ventured into genealogical territory beyond the African American community, featuring celebrities like Yo-Yo Ma, Eva Longoria, Louise Erdrich, and Meryl Streep. Gates will launch a television series on genetic genealogy in 2012.

67. The genomics of race was also used to articulate the correct way to understand race in *Race, the Power of an Illusion* (DVD), Larry Adelman, producer (PBS, 2003). The four-part series opened with an episode on the power of DNA to show students their true ancestry.

68. On semantic configurations, see Didier Fassin, "The Embodied Past: From Paranoid Style to Politics of Memory in South Africa," *Social Anthropology* 16 (2008), 312–28.

69. Gates is also a board member of AesRx, a pharmacogenomics company that develops therapies for sickle cell anemia.

70. In 2007 the genealogy company Family Tree DNA partnered with Gates to create AfricanDNA. AfricanDNA combines genomic and historical genealogical methods to help African Americans find their ancestral roots. Harvard University's Personal Genomes Project also recruited Gates and his father to be the first African Americans to have their whole genomes analyzed. In 2010 Gates joined the board of AesRx, where he will guide the company in its efforts to create the first pharmacogenomic drug targeted at African Americans. His reputation has lent credence to genomics' race-positive orientation.

71. http://www.africandna.com/history.aspx.

72. http://www.oxfordancestors.com.

73. Bryan Sykes, *The Seven Daughters of Eve: The Science That Reveals Our Genetic Ancestry* (New York: Norton, 2002); *Blood of the Isles: Exploring the Genetic Roots of Our Tribal History* (New York: Bantam Books, 2006).

74. http://www.africanancestry.com.

75. Scientists involved in the ancestry industry learn to maneuver between scientific fact, historical fiction, and anecdotal fantasy. They understand that they must bring results to life with meaningful story. This is true even in situations where scientists are reporting on maternal and paternal lineages that cover less than one percent of a person's ancestral heritage.

For a critique of this practice, see Deborah A. Bolnick, Duana Fullwiley, Troy Duster, Richard S. Cooper, et al., "The Science and Business of Genetic Ancestry Testing," *Science* 318 (2007), 399–400.

76. Jonathan Kahn, "Race in a Bottle," *Scientific American* (Aug. 2007); Epstein, *Inclusion*.

77. BBC, *Motherland*.

78. Nelson, "Factness of Diaspora."

79. Mark developed a critical stance toward assuming a solely genetic identity when his own enthusiasm to adopt a genetic identity led him to blindly affiliate with a pro-slavery group he knew little about.

80. On racial representation and signification, see the work of Stuart Hall and Judith Butler.

81. Roberts warns of the inadequacy of a genomics-based race to work with social policies designed to redress social inequality. Dorothy Roberts, "Race and the New Biocitizen," in *What's the Use of Race? Modern Governance and the Biology of Difference*, edited by Ian Whitmarsh and David S. Jones (Cambridge, MA: MIT Press, 2010).

82. Tallbear, "Native American DNA."

83. Kimberly Tallbear and Deborah Bolnick, "'Native American DNA' Tests." Tallbear and Bolnick also say that for tribes that administer benefits to members, genealogy tests are believed to hold the potential for shortening tribal rosters.

84. Reardon has presented some of the problems with the way scientists have moved toward an earlier public engagement and ethical framing for research in light of asymmetries between lay and scientific experts. Reardon, "Finding Oprah's Roots, Losing the World: Beyond the Liberal Anti-Racist Genome" (2009), http://globetrotter .berkeley.edu/bwep/colloquium/papers/FindingOprahLosingWorld%20Oct%203%20 2009%20PDF.pdf.

85. See Sheldon Krimsky and Tania Simoncelli, *Genetic Justice: DNA Data Banks, Criminal Investigations, and Civil Liberties* (New York: Columbia University Press, 2011); Pamela Sankar, "Forensic DNA Phenotyping: Reinforcing Race in Law Enforcement," in *What's the Use of Race? Modern Governance and the Biology of Difference*, edited by Ian Whitmarsh and David S. Jones (Cambridge, MA: MIT Press, 2010); and Jonathan Kahn, "Race, Genes and Justice: A Call to Reform the Presentation of Forensic DNA Evidence in Criminal Trials," *Brooklyn Law Review* 74 (2009), 325.

Chapter 2

1. Here I am obliquely referring to Emile Durkheim's concept of "collective effervescence": an energy that accrues from a gathering of people but is greater than the sum of its parts. Durkheim, *The Elementary Forms of the Religious Life*, translated by Carol Cosman and Mark Sydney Cladis (Oxford, UK: Oxford University Press, 2001).

2. See Loet Leydesdorff and Martin Meyer, "Triple Helix Indicators of Knowledge-Based Innovation Systems," *Research Policy* 35 (2006), 1441–49; Loet Leydesdorff and Henry Etzkowitz, "Can 'The Public' Be Considered as a Fourth Helix in University-Industry-Government Relations?" Report of the Fourth Triple Helix Conference, *Science and Public Policy* 30 (2003), 55–61; and Maria Häyrinen-Alestalo and Ulla Peltalo, "The Problem of a Market-Oriented University," *Higher Education* 52 (2006), 251–81.

3. To this model I would add the nonprofit organizations that are often dually funded by government and private foundations.

4. For histories of Celera, its parent companies, and its role in the Human Genome Project, see Paul Rabinow and Talia Dan-Cohen, *A Machine to Make a Future: Biotech Chronicles* (Princeton, NJ: Princeton University Press, 2005); and James Shreeve, *The Genome War: How Craig Venter Tried to Capture the Code of Life and Save the World* (New York: Alfred A. Knopf, 2004).

5. For more on CEPH's relationship to the early Human Genome Project, see Rabinow and Dan-Cohen, *Machine to Make a Future*.

6. Pierre Bourdieu, *Pascalian Meditations* (Stanford, CA: Stanford University Press, 1997); *Science of Science and Reflexivity*, translated by Richard Nice (Chicago: University of Chicago Press, 2004).

7. This first attempt to sequence and map the entire 3.2 billion nucleotide base pairs of all DNA belonging to the human organism stemmed from a technological boom in the fields of informatics and genetics. Some important precursors were the invention of the polymerase chain reaction, microcomputers, and the establishment of GenBank at Los Alamos National Laboratory. Robert Mullan Cook-Deegan, "Origins of the Human Genome Project," *Risk* 5 (1994), http://www.fplc.edu/risk/vol5/spring/cookdeeg.htm. For a historical look at the complete project, see Victor K. McElheny, *Drawing the Map of Life: Inside the Human Genome Project* (New York: Perseus, 2010).

8. For more on the role of the Department of Energy and its prior gene databases at Los Alamos, Lawrence Livermore, and Oak Ridge National Laboratories, see Charles R. Cantor, "Orchestrating the Human Genome Project," *Science* 248 (1990), 49–51.

9. Francis Collins, Michael Morgan, and Aristides Patrinos, "The Human Genome Project: Lessons from Large-Scale Biology," *Science* 300 (2003), 286–90.

10. John Burris, Robert Cook-Deegan, and Bruce Alberts, "The Human Genome Project After a Decade," *Nature Genetics* 20 (1998), 333–35.

11. Of the initial 220 members, 103 were from the United States, 33 were from the United Kingdom, 15 were from France, and no more than 11 were from Australia, Austria, Belgium, Canada, Denmark, East Germany, Finland, Greece, Israel, Iceland, Italy, Japan, Norway, South Africa, Spain, Sweden, Switzerland, Netherlands, Soviet Union, and West Germany (most contributing a mere one or two members).

12. The primary sequencing centers were U.S. Department of Energy Joint Genome Institute, Walnut Creek, California; Baylor College of Medicine Human Genome Sequencing Center, Department of Molecular and Human Genetics, Houston, Texas; The Wellcome Trust Sanger Institute, The Wellcome Trust Genome Campus, Hinxton, Cambridgeshire, U.K.; Washington University School of Medicine Genome Sequencing Center, St. Louis, Missouri; and Whitehead Institute/MIT Center for Genome Research, Cambridge, Massachusetts. U.S. Department of Energy (DOE), "Human Genome Project," in *Human Genome Research Sites*, http://www.ornl.gov/sci/techresources/Human_Genome/research/centers.shtml.

13. Epstein, *Inclusion*, 4–5, 46, 54, 75.

14. *Ibid.*, 44–45.

15. Reardon, *Race to the Finish*, 23–26, 46.

16. An autosome is a chromosome that is not one of the two sex chromosomes—each individual possesses a paternal and maternal copy of the twenty-two human autosomes plus two sex chromosomes. Thus the Genome Project sequence would be a map of half of the DNA in the nucleus of one individual.

17. Pierre Bourdieu discusses autonomization in terms of the process by which fields vie for an independent space to develop their own rules of struggle, domination, and legitimacy. Though Bourdieu is referring to abstract fields such as art, science, and the economy, the struggle to develop unique configurations of capital can be seen in the development of new scientific fields. Bourdieu, *The Logic of Practice* (Stanford, CA: Stanford University Press, 1990).

18. NIH, *Understanding Our Genetic Inheritance: The U.S. Human Genome Project. The First Five Years: Fiscal Years 1991–1995* (Washington, DC: NIH Publication No. 90-1590, 1990).

19. Francis Collins and David Galas, "A New Five-Year Plan for the U.S. Human Genome Project," *Science* 262 (1993), 46.

20. Elke Jordan, "The Human Genome Project: Where Did It Come From, Where Is It Going?" *American Journal of Human Genetics* 51 (1992), 4.

21. Fatimah Jackson, "Scientific Implications and Ethical Ramifications of a Non-Representative Human Genome Project: African American Response," *Science and Engineering Ethics* 4 (1998), 157.

22. Epstein, *Inclusion*, 17.

23. Omi and Winant, *Racial Formations*, 164. Here I am using the term more broadly than Michael Omi and Howard Winant, who emphasize the domain of social theory. However, I am inspired by their expansion of the term to spheres beyond governance and science.

24. *Ibid.*, 56. A central theme in my argument is that in order to treat science and politics symmetrically, one must apply the tried and tested theories of our time to scientists themselves. When racial formations theory is applied to genomicists, the political goals inherent in their basic research strategies become apparent.

25. Epstein, *Inclusion*, 59, 73–79. Epstein provides insightful analysis of the debates and "false starts" regarding the inclusion-and-difference policy paradigm.

26. See "The New Face of America," *Time* (Nov. 18, 1993).

27. David Hollinger, *Postethnic America: Beyond Multiculturalism* (New York: Basic Books, 2006).

28. Postracialism is the doctrine that society is beyond racial inequality, and therefore it need not address racism with policy, resources, or dialogue. See Chapter 7 for a discussion of contemporary postracialist discourse.

29. U.S. Department of Health and Human Services (HHS), *Healthy People 2000: Our Nation's Prevention Agenda* (1994), http://www.crisny.org/health/us/health7.html.

30. Leslie Roberts and Ann Gibbons, "A Genetic Survey of Vanishing Peoples: Racing the Clock, Two Leaders in Genetics and Evolution Are Calling for an Urgent Effort to Collect DNA From Rapidly Disappearing Indigenous Populations," *Science* 252 (1991), 1615. Bodmer was a longtime collaborator with Luca Cavalli-Sforza, having coauthored *The Genetics of Human Populations* (Mineola, NY: Dover, 1999 [1971]).

31. For an excellent analysis of project methods in relation to broader anthropological debates, see Reardon, *Race to the Finish*.

32. Leslie Roberts, "How to Sample the World's Genetic Diversity," *Science* 257 (1992), 1205.

33. Recall the quadrilateral nature of minority inclusion and racial politics. A protocol dedicated to ethnic inclusion, or inclusion based on cultural and genealogical similarity, would not necessarily impact the histories of discrimination that characterize the racial experience.

34. Roberts, "How to Sample the World's Genetic Diversity," 1204; Anne Bowcock and Luca Cavalli-Sforza, "The Study of Variation in the Human Genome," *Genomics* 11 (1991), 491.

35. Roberts, "How to Sample the World's Genetic Diversity," 1204.

36. Bowcock and Cavalli-Sforza, "Study of Variation in the Human Genome," 490.

37. *The NIH Revitalization Act of 1993* (PL 103-43, 1993), http://grants.nih.gov/grants/guide/notice-files/not94-100.html; NIH, *Guidelines on the Inclusion of Women and Minorities as Subjects in Clinical Research* (1994), http://grants.nih.gov/grants/guide/notice-files/not94-100.html. As Snait Gissis has shown, *Revitalization* replaced the previous policy language of encouragement with that of requirement. Gissis, "When Is 'Race' a Race?" 443.

38. At this point, ELSI also turned to study race in establishing two studies on race and population representation. One, conducted by sociologist Troy Duster, explored how families at risk for three common recessive diseases popularly understood to be restricted to particular races—sickle cell anemia, thalassemia, and cystic fibrosis—experienced genetic testing. The other, led by pharmacologist Ralph Trottier, explored African American experiences of genetic testing, screening, and counseling patients around sickle cell anemia. Though these studies advanced the field toward an interest in race, genomicists were not responsible for these studies and they did not directly impact the sampling protocols of the field. Neither study directly explored the impact genomic advances would have on the idea of race itself, much less the relevance of sampling representation. Human Genome Project press releases still listed race-free ELSI priorities

as their key problems. Michael M. Gottesman and Francis S. Collins, "The Role of the Human Genome Project in Disease Prevention," *Preventive Medicine* 23 (1994), 591–94; Mark S. Guyer and Francis S. Collins, "How Is the Human Genome Project Doing, and What Have We Learned So Far?" *Proceedings of the National Academy of Sciences* 92 (1995), 10841–48.

39. Richard J. Herrnstein and Charles A. Murray, *The Bell Curve: Intelligence and Class Structure in American Life* (New York: Simon and Schuster, 1996).

40. Human Genome Diversity Project, address delivered by Luca Cavalli-Sforza, Stanford University, to a special meeting of UNESCO (Paris, Sept. 21, 1994).

41. Timothy Rebbeck and Pamela Sankar define "minor ethnicity" as ethnicity beneath the level of racial and national identity. Timothy R. Rebbeck and Pamela Sankar, "Ethnicity, Ancestry, and Race in Molecular Epidemiologic Research," *Cancer Epidemiology Biomarkers and Prevention* 14 (2005), 2468–71.

42. Reardon, *Race to the Finish,* 5.

43. David Dickson, "Whose Genes Are They Anyway?" *Nature* 381 (1996), 11–14.

44. Reardon, *Race to the Finish,* 51.

45. William J. Schull, correspondence, *Nature* 390 (1997), 221.

46. Gannett, "Racism and Human Genome Diversity Research," S478.

47. The three goals are "ensuring that the initial version of the complete human DNA sequence is derived from multiple donors," "providing donors with the opportunity to make an informed decision about whether to contribute their DNA to this project," and "taking effective steps to ensure the privacy and confidentiality of donors." U.S. Department of Energy (DOE), *To Know Ourselves* (1996), http://www.ornl.gov/sci/ techresources/Human_Genome/publicat/tko/index.html.

48. *Ibid.*

49. For analysis of the Tuskegee Syphilis Study and its social implications, see Susan Reverby, *Examining Tuskegee: The Infamous Syphilis Study and Its Legacy* (Chapel Hill: University of North Carolina Press, 2009).

50. Francis S. Collins, Lisa D. Brooks, and Aravinda Chakravarti, "A DNA Polymorphism Discovery Resource for Research on Human Genetic Variation," *Genome Research* 8 (1998), 1229–31.

51. For a more skeptical assessment of the Discovery Resource's retreat from indigenous communities, see Reardon, *Race to the Finish,* 157. Reardon worried that scientists might lose confidence in their ability to manage the racial landscape. However, I find that scientists see these moments of failure as turning points, motivational moments for refashioning the field's ethical strategy.

52. U.S. Department of Energy (DOE), "A DNA Polymorphism Discovery Resource," *Human Genome News* 10 (Feb. 1999), http://www.genome.gov/pfv.cfm?pageID =10001552. This statement has been edited for the current Discovery Resource webpage of the Genome Research Institute. It now reads: "the resource includes samples representative of the genetic diversity found in the U.S. population."

53. Elliott Marshall, "'Playing Chicken' over Gene Markers," *Science* 278 (1997), 2047.

54. For more on the competition, see U.S. Department of Energy (DOE), "The

Human Genome Project and the Private Sector: A Working Partnership," http://www .ornl.gov/sci/techresources/Human_Genome/project/privatesector.shtml; and Elliot Marshall, "Sharing the Glory, Not the Credit," *Science* 291 (2001), 1189–93.

55. Venter's comments linking gender representation and racial representation show that the inclusion-and-difference paradigm described by Epstein was a dominant framework for scientists working at the turn of the century. In Celera's portion of the project, like the government-led portion, samples were disidentified and pooled. However, the fact of diversity and racial representation was integral to the foundations of Celera's sampling protocol.

56. Venter reported in *Science*:

Prospective donors were asked, on a voluntary basis, to self-designate an ethnogeographic category (e.g., African-American, Chinese, Hispanic, Caucasian, etc.). We enrolled 21 donors. . . . DNA from five subjects was selected for genomic DNA sequencing: two males and three females—one African-American, one Asian-Chinese, one Hispanic-Mexican, and two Caucasians. The decision of whose DNA to sequence was based on a complex mix of factors, including the goal of achieving diversity as well as technical issues such as the quality of the DNA libraries and availability of immortalized cell lines. (J. Craig Venter, Mark D. Adams, Eugene W. Myers, Peter W. Li, et al., "The Sequence of the Human Genome," *Science* 291 [2001], 1306)

57. For a treatment of the literature on the interinstitutional "thickening" of paradigms, see Epstein, *Inclusion*, 117.

58. Francis S. Collins, Ari Patrinos, Elke Jordan, Aravinda Chakravarti, et al., "New Goals for the U.S. Human Genome Project: 1998–2003," *Science* 282 (1998), 682–83.

59. *Ibid.*, 688.

60. NIH, *Studies of the Ethical, Legal and Social Implications of Research into Human Genetic Variation*, RFA: HG-99-002. The first projects funded included: Sandra Lee's "The Ethics of Identifying Race in the New Genetics"; Howard Markel's "The Stigma of Disease: Implications for Testing"; and David Micklos's "Digital Image Archive on the American Eugenics Movement" (cf. Stevens, "Racial Meanings and Scientific Methods").

61. Office of the Press Secretary, The White House, remarks by the President, Prime Minister Tony Blair of England (via satellite), Dr. Francis Collins, Director of the National Human Genome Research Institute, and Dr. Craig Venter, President and Chief Scientific Officer, Celera Genomics Corporation, on the completion of the first survey of the entire Human Genome Project (The East Room, June 26, 2000; 10:19 A.M. EDT).

62. Clinton also benefited from these remarks in that genomics served to justify his political message of equality.

63. Office of the Press Secretary, The White House, remarks.

64. Angier, "Do Races Differ? Not Really, Genes Show."

65. Marcus Gee, "Debunking the Myth of Race," *Globe and Mail* (Feb. 15, 2000), A13.

66. For a review of pharmacogenomics in the early 2000s, see Hong-Guang Xie, Richard B. Kim, Alastair J.J. Wood, and C. Michael Stein, "Molecular Basis of Ethnic Difference in Drug Disposition and Response," *Annual Review of Pharmacology and Toxi-*

cology 41 (2001), 815–50; and Kathryn A. Phillips, David L. Veneestra, Eyal Oren, Jane K. Lee, et al., "Potential Role of Pharmacogenomics in Reducing Adverse Drug Reactions," *Journal of the American Medical Association* 286 (2001), 2270–79.

67. Of particular concern were the cytochrome P450 oxidases and thiopurine methyltransferases. See Xie, Kim, Wood, and Stein, "Molecular Basis of Ethnic Difference in Drug Disposition and Response."

68. Rick Weiss, "The Promise of Precision Prescriptions: 'Pharmacogenomics' also Raises Issues of Race, Privacy," *Washington Post* (June 24, 2000), A01.

69. Derek V. Exner, Daniel L. Dries, Michael J. Domanksi, and Jay N. Cohn, "Lesser Response to Angiotensin-Converting-Enzyme Inhibitor Therapy in Blacks as Compared with White Patients with Left Ventricular Dysfunction," *New England Journal of Medicine* 344 (2001), 1351–57.

70. Clyde W. Yancy, Michael B. Fowler, Wilson S. Colucci, Edward M. Gilbert, et al., "Race and the Response to Andrenergic Blockade with Carvedilol in Patients with Chronic Heart Failure," *New England Journal of Medicine* 344 (2001), 1358–65.

71. *Ibid.*; Exner, Dries, Domanksi, and Cohn, "Lesser Response to Angiotensin-Converting-Enzyme Inhibitor Therapy in Blacks as Compared with White Patients with Left Ventricular Dysfunction." "Black" and "white" are the *New England Journal of Medicine*'s standard terms. In a 2008 conference on race, genetics, and governance, the *Journal of the American Medical Association*'s deputy editor, Margaret Winker, expressed the journal's intent on adopting these color-coded standards. "What's the Use of Race?" MIT, Apr. 25–26, 2008.

72. Robert Schwartz, "Racial Profiling in Medical Research," *New England Journal of Medicine* 344 (2001), 1393.

73. Alastair J.J. Wood, "Racial Differences in Response to Drugs—Pointers to Genetic Differences," *New England Journal of Medicine* 344 (2001), 1395. Wood coauthored the Xie, Kim, Wood, and Stein (2001) review designed to establish drug response gene frequencies as ethnically variant.

74. Schwartz, "Racial Profiling in Medical Research."

75. This debate continued in the maiden volume of the *Pharmacogenomics Journal* (2001), where legal scholars, bioethicists, and genomicists weighed in on whether pharmacogenomics should use population-based data at all. University of Cincinnati scientists Daniel Nebert and Anil Menon said that genomics needed more genotypic information that "should be accomplished by mechanisms based on scientific reason, rather than mandates for 'racial inclusion'" but also agreed with interlocutors Werner Kalow, Vural Ozdemir, and Laszlo Tothfalusi that "the inclusion of an ethnic minority in a clinical study may reveal information on the frequency of an unknown genetic variant which affects a drug response; such information can be of clinical importance and may increase the safety of some members of that ethnic population." Like the antihypertensive trial reports, their arguments linked the fate of pharmacogenomics to choices between greater genomic acuity and ethical inclusion.

76. CNN, "Should Doctors Consider Skin Color When Prescribing Drugs?" *CNN Live at Daybreak* (May 3, 2001).

77. NPR, "New Study Raises Questions About When It's Appropriate to Focus on Race in Medical Research," *All Things Considered* (May 2, 2001).

78. Stolberg, "Shouldn't a Pill Be Colorblind?"

79. See Reardon, *Race to the Finish*, 53, for analysis of the way Diversity Project scientists manipulated this term.

80. Jonathan Leake, "U.S. Nobel Laureate Sparks Sex-Race Flap—DNA Hero Provokes Academics by Linking Libido to Skin Colour," *Toronto Star* (Jan. 7, 2001).

81. Thomas Murray, "Race, Ethnicity, and Science: The Haplotype Genome Project," *Hastings Center Report* 31 (2001), 7.

82. Wade, "For Genome Mappers, the Tricky Terrain of Race Requires Some Careful Navigating."

83. Phillips, Veenstra, Oren, Lee, and Sadee, "Potential Role of Pharmacogenomics in Reducing Adverse Drug Reactions"; Associated Press, "Potential of Tailoring Drugs to Genetic Makeup Confirmed—But Challenges Remain" (Nov. 13, 2001).

84. Chandler, "Heredity Study Eyes European Origins," A22.

85. Likewise, in the immediate scientific realm, methodological and technological advances precipitated a population structure focus. Research also continued to elucidate the patterns of linkage disequilibrium: Ning Yu, Feng-Chi Chen, Satoshi Ota, Lynn B. Jorde, et al., "Larger Genetic Differences Within Africans Than Between Africans and Eurasians," *Genetics* 161 (2002), 269–74; and David Reich, Stephen F. Schaffner, Mark J. Daly, Gil McVean, et al., "Human Genome Sequence Variation and the Influence of Gene History, Mutation and Recombination," *Nature Genetics* 32 (2002), 135–42. Other studies looked at admixture linkage disequilibrium in the human genome: Heather E. Collins-Schramm, Carolyn M. Phillips, Darwin J. Operario, Jane S. Lee, et al., "Ethnic-Difference Markers for Use in Mapping by Admixture Linkage Disequilibrium," *American Journal of Human Genetics* 70 (2002), 737–50; and Carrie L. Pfaff, Rick A. Kittles, and Mark D. Shriver, "Adjusting for Population Structure in Admixed Populations," *Genetic Epidemiology* 22 (2002), 196–201. Of particular import was the discovery that haplotypes, segments of linked DNA, exist throughout the genome and are its signature pattern: Goldstein, "Islands of Linkage Disequilibrium." Capitalizing on the increasingly evident simplicity of the genome, scientists began searching for markers that could represent population-specific hereditary material: Collins-Schramm, Phillips, Operario, Lee, et al., "Ethnic-Difference Markers for Use in Mapping by Admixture Linkage Disequilibrium"; and David E. Reich and David B. Goldstein, "Detecting Association in a Case-Control Study While Correcting for Population Stratification," *Genetic Epidemiology* 20 (2002), 4–16. Detailing population variation became the field's priority.

86. NIH, *Report of the First Community Consultation on the Responsible Collection and Use of Samples for Genetic Research* (Sept. 25–26, 2000), http://www.nigms.nih.gov/ News/Reports/community_consultation.htm.

87. U.S. Department of Health and Human Services (HHS), *Healthy People 2010: Understanding and Improving Health*, 2nd edition (Washington, DC: U.S. Government Printing Office, Nov. 2000); *Tracking Healthy People 2010*, http://www.healthypeople .gov/Document/tableofcontents.htm.

88. U.S. Department of Health and Human Services (HHS), *The National Human Genome Research Institute Strategic Plan for Reducing Health Disparities: Fiscal Years 2002–2006: Draft October 6, 2000*, http://www.nih.gov/about/hd/strategicplan.pdf.

89. Nancy Kreiger showed that from 1995 to 2004, the institutes of the NIH announced 181 new grants indexed by race and genetics. My review of the Strategic Plans for Eliminating Health Disparities of each institute confirmed this finding. Kreiger, "Stormy Weather: Race, Gene Expression, and the Science of Health Disparities," *American Journal of Public Health* 95 (2005), 2155–60.

90. Participating companies were APBiotech, AstraZeneca Group PLC, Aventis, Bayer Group AG, Bristol-Myers Squibb Co., F. Hoffmann-La Roche, Glaxo Wellcome PLC, IBM, Motorola, Novartis AG, Pfizer Inc., Searle, and SmithKline Beecham PLC. TSC-funded labs were the Whitehead Institute, Sanger Centre, Washington University, and Stanford University. Data were processed at Cold Spring Harbor Laboratory. Funding totaled over $60 million covering a two-year period.

91. U.S. Department of Energy, *SNP Fact Sheet*, http://www.ornl.gov/sci/tech resources/Human Genome/faq/snps.shtml#snps.

92. U.S. Department of Energy, *Human Genome Project and SNP Consortium Announce Collaboration to Identify New Genetic Markers for Disease*, http://www.genome .gov/10001456.

93. International SNP Map Working Group, "A Map of Human Genome Sequence Variation Containing 1.42 Million Single Nucleotide Polymorphisms," *Nature* 409 (2001), 928–33.

94. Elizabeth Pennisi, "So Many Choices, So Little Money," *Science* 294 (2001), 82–85.

95. Nila Patil, Anthony J. Berno, David A. Hinds, Wade A. Barett, et al., "Blocks of Limited Haplotype Diversity Revealed by High Resolution Scanning of Human Chromosome 21," *Science* 294 (2001), 1719–23.

96. Gudmundur A. Thorisson and Lincoln D. Stein, "The SNP Consortium Website: Past, Present and Future," *Nucleic Acids Research* 31 (2003), 124–27.

97. Eliot Marshall, "DNA Studies Challenge the Meaning of Race," *Science* 282 (1998), 654–55; "'Playing Chicken' over Gene Markers."

98. Will Dunham, "Human Genetic Variation Found in Human Beings," Reuters News (July 12, 2001).

99. Amid planning, the NIH and FDA reinforced their inclusion policies. The Genome Research Institute also commenced talks to further institutionalize a Minority Action Plan that would recruit minority researchers for ELSI and genomic research, and fund new models for minority-focused genomic research. NIH, *NIH Policy on Reporting Race and Ethnicity Data: Subjects in Clinical Research*, http://grants.nih.gov/grants/ guide/notice-files/not94-100.html; FDA, "Draft Guidance for Industry on the Collection of Race and Ethnicity Data in Clinical Trials for FDA Regulated Products; Availability," *Federal Register* 68, (2003); U.S. Department of Health and Human Services (HHS), *Guidance for Industry: Collection of Race and Ethnicity Data in Clinical Trials*, http:// www.fda.gov/cber/gdlns/racethclin.htm.

100. Also see Reardon, "Finding Oprah's Roots, Losing the World." On how these

talks relate to the emergence of personal genomics, see Reardon, "'Persons' and 'Genomics' of Personal Genomics," 97.

101. See http://www.westafricanbioethics.net/aba-alamu for a photograph of the enrollment of Yoruba from Ibadan, Nigeria, in which community leaders hold framed certificates.

102. Altshuler is now the director of medical and population genetics at the Broad and a chief planner of the 1000 Genomes Project—which is sequencing the entire genomes of one thousand HapMap samples. His participation ensures that this cultural competence stance will continue to influence the sampling protocols of global genomics.

103. International HapMap Consortium, "The International HapMap Project," *Nature* 426 (2003), 791.

104. International HapMap Project, *Guidelines for Referring to the HapMap Populations in Publications and Presentations*, http://www.hapmap.org/citinghapmap.html. The complete recommended language for naming the populations included in the HapMap (which reflects both the ancestral geography of each population and the geographic location where the samples from that population were collected) is: Yoruba in Ibadan, Nigeria (abbreviation: YRI); Japanese in Tokyo, Japan (abbreviation: JPT); Han Chinese in Beijing, China (abbreviation: CHB); CEPH (Utah residents with ancestry from northern and western Europe) (abbreviation: CEU). After the complete descriptor for a population has been provided, it is acceptable to use a shorthand label for that population (e.g., "Yoruba," "Japanese," "Han Chinese," "CEPH") or the abbreviation for that population (e.g., "YRI," "JPT," "CHB," "CEU") in the remainder of the article or presentation. However, the full descriptor for each population should be provided before such shorthand labels are used. This will help to avoid the risks associated with overgeneralization of findings.

105. *Ibid.*

106. According to Reardon, "the only thing 'the community' chose was that in addition to English, their name would be printed in their native language." Reardon, "Finding Oprah's Roots, Losing the World," 8.

107. http://www.genomecenter.howard.edu.

108. Aristides Patrinos, "'Race' and the Human Genome," *Nature Genetics* 36 (2004), S1.

109. The Genome Research Institute was also further institutionalizing health disparities research and a minority health program at this time. In the institute's "New NHGRI Vision for Genomics," one of the ten Grand Challenges the institute set for itself was "Understand the relationships among genomics, race, and ethnicity, and the consequences of uncovering these relationships."

110. Avram Goldstein and Rick Weiss, "Howard U Plans Genetics Database," *Washington Post* (May 28, 2003), A06.

111. A lone voice of dissent, Troy Duster warned the public that race would likely be reified at the genetic level by such an endeavor. He charged scientists to focus on environmental causes: "Do [patients] live near toxic waste dumps? Do they have more stress problems and hypertension?" *Ibid.*

112. Anne L. Taylor, Susan Ziesche, Clyde Yancy, Peter Carson, et al., for the African-American Heart Failure Trial Investigators, "Combination of Isosorbide Dinitrate and Hydralazine in Blacks with Heart Failure," *New England Journal of Medicine* 351 (2004), 2055.

113. *Ibid.*

114. In an accompanying critique, however, policy analyst and law professor Gregg Bloche contended there was no financial incentive for the pharmaceutical company producing the drug to do further research, once the drug was approved. In fact, since the drug was already a rerelease of a therapy whose patent was about to expire, there would be no legal responsibility to conduct another trial until its next expiration in 2020. M. Gregg Bloche, "Race-Based Therapeutics," *New England Journal of Medicine* 351 (2004), 2035–37.

115. The drug's producer, Nitromed, has enlisted other black health advocates such as Gary Puckrein, executive director of the National Minority Health Month Foundation in Washington, DC. See Gary Puckrein, "BiDil: From Another Vantage Point," *Health Affairs* 25 (2005), w368–w374; and Gary Puckrein and Clyde W. Yancy, "BiDil's Impact," *Nature Biotechnology* 23 (2005), 1343.

116. David A. Hinds, Laura L. Stuve, Geoffrey B. Nilsen, Eran Halperin, et al., "Whole-Genome Patterns of Common DNA Variation in Three Human Populations," *Science* 307 (2005), 1073; *HapMap News* 2 (Camden, NJ: Coriell Institute for Medical Research, 2004).

117. Anna C. Need and David B. Goldstein, "Genome-wide Tagging for Everyone," *Nature Genetics* 38 (2006), 1227–28; Cha Tian, David A. Hinds, Russell Shigeta, Rick Kittles, et al., "A Genomewide Single-Nucleotide-Polymorphism Panel with High Ancestry Information for African American Admixture Mapping," *American Journal of Human Genetics* 79 (2006), 640–49.

Mary-Anne Enoch, Pei-Hong Shen, Ke Xu, Colin Hogdkinson, et al., "Using Ancestry-Informative Markers to Define Populations and Detect Population Stratification," *Journal of Psychopharmacology* 20 (2006), S19–S26.

118. Anna Helgadottir, Ancrei Manolescu, Agnar Helgason, Gumar Thorliefson, et al., "A Variant of the Gene Encoding Leukotriene A4 Hydrolase Confers Ethnicity-Specific Risk of Myocardial Infarction," *Nature Genetics* 38 (2006), 68–74; R. Huang and S. E. Poduslo, "CYP19 Haplotypes Increase Risk for Alzheimer's Disease," *Journal of Medical Genetics* 43 (2006), e42.

119. For a review, see Pardis C. Sabeti, Patrick Varilly, Ben Fry, Jason Lohmueller, et al. and the International HapMap Consortium, "Genome-wide Detection and Characterization of Positive Selection in Human Populations," *Nature* 449 (2007), 913–19.

120. Kahn, "Beyond BiDil," 82.

121. This is not to mention pharmaceutical and biotechnology companies that are also increasingly marketing their diagnostics directly to the consumer. See, for example, Pharmigene's plea for a Southeast Asian diagnostic for the psychotropic drug Tegretol, http://www.genomeweb.com/node/948605.

122. Kahn, "Beyond BiDil," 74.

123. https://genographic.nationalgeographic.com/genographic/lan/en/faqs_about .html#Q1.

124. Though several indigenous and first peoples organizations, including the United Nations Permanent Forum on Indigenous Issues, issued a stop-and-desist recommendation to the project in 2006, numerous tribes from all continents agreed to take part. As of the project's closing date, over fifty thousand tribes were sampled.

125. http://www.genome.gov/Pages/Newsroom/Webcasts/Transcript-H3-Africa PressConference.pdf.

Chapter 3

1. Melinda Deslatte, "DNA Test Pointed to Black Man in Louisiana Serial Killer Case," Associated Press (June 4, 2003).

2. Nicholas Wade, "Unusual Sse of DNA," *New York Times* (June 3, 2003).

3. Deslatte, "DNA Test Pointed to Black Man in Louisiana Serial Killer Case."

4. Interest in eliciting the hereditary characteristics of race was so widely accepted that by 2003 it permeated biomedicine. For example, the *Journal of the American Academy of Dermatology* (June) and *Dermatologic Clinics* (October) each hosted their first issues on "Ethnic Skin and Hair." The publication of studies stratifying analyses by race and ethnicity was the springboard for further issues and articles.

5. Oren Dorell, "Police Enlisting DNA Tool," *News and Observer* [Raleigh, NC] (Aug. 1, 2003); Marcos Mocine-McQueen, "DNA Test Suggests Race of Woman's Killer in '97 [Corrected 01/23/04]," *Denver Post* (Jan. 21, 2004), B01.

6. See, for example, Mildred K. Cho and Pamela Sankar, "Forensic Genetics and Ethical, Legal and Social Implications Beyond the Clinic," *Nature Genetics* 36 (2004), S8–S12; "In Reply to 'Getting the Science and the Ethics Right in Forensic Genetics,'" *Nature Genetics* 37 (2005), 450–51; and Ossorio, "About Face."

7. Catherine Bliss, "Mapping Admixture by Race," *International Journal of Technology, Knowledge and Society* 4 (2008), 79–83.

8. Under a continental framework, admixture mapping has been used to better understand asthma, cancer, cardiovascular disease, diabetes, hypertension, multiple sclerosis, osteoporosis, and systemic lupus; track trait prevalence related to adiposity, body mass, depression, obsessive compulsive disorder, personality traits, pigmentation, pregnancy disorders, and the pharmacogenetics of alcohol and drugs; admixture mapping has found candidate genes for cardiovascular disease, hypertension, multiple sclerosis, and prostate cancer; and admixture mapping has been recommended for use in epilepsy, inflammatory bowel disorder, nephropathy, aging, cleft palate, and eating disorders.

9. Elad Ziv and Esteban Gonzalez Burchard, "Human Population Structure and Genetic Association Studies," *Pharmacogenomics* 4 (2003), 431–41; Hui-Ju Tsai, Shweta Choudhry, Mariam Naqvi, William Rodriguez-Cintron, Esteban Gonzalez Burchard, and Elad Ziv, "Comparison of Three Methods to Estimate Genetic Ancestry and Control for Stratification in Genetic Association Studies Among Admixed Populations," *Human Genetics* 118 (2005), 424–33.

10. CBS News, "DNA Testing for Sale" (Feb. 28, 2003); Dana Hawkins Simons, "Getting DNA to Bear Witness," *US News* (June 23, 2003); Malcolm Ritter, "As Role of Race in Research Questioned, a DNA Databank Proposed for Blacks," Associated Press (July 20,

2003); Nicholas Wade, "Is Race Real?" *New York Times* (July 11, 2003), 17; Mark D. Shriver and Rick A. Kittles, "Genetic Ancestry and the Search for Personalized Genetic Histories," *Nature Reviews Genetics* 5 (2004), 616; and Mark Shriver, Tony Frudakis, and Bruce Budowle, "Getting the Science and the Ethics Right in Forensic Genetics," *Nature Genetics* 37 (2005), 449–51.

11. One can find original company biographies in the annals of business databases that state DNAPrint's core purpose to be racial ancestry analysis. In 2008 the company reinserted a definition of race into their online glossary. The full text is in Note 13.

12. "Forensic Flyer: DNAWitness 2.5," http://www.dnaprint.com.

13. The full text read:

Race in general usage includes both a cultural and biological feature of a person or group of people. Given the fact that physical differences between populations are often accompanied by cultural differences, it has been difficult to separate these two elements of race. Over the past few decades there has been a movement in several fields of science to oversimplify the issue declaring that race is "merely a social construct." While, indeed this may often be true, depending on what aspect of variation between people one is considering, it is also true that there are biological differences between the populations of the world. One clear example of a biological difference is skin color. There is a strong genetic component to the level of pigmentation in a person's skin and there are dramatic differences across populations. Pigmentation is, however, only skin deep and really a simple heritable trait in light of the complex environments in which we all live and how these environments affect our individual and group quality of life is far beyond our ability to understand as scientists.

The biological feature of race is largely based on the genetic structure of human populations. This structure is a nested hierarchy from East to West where populations in the Americas and the South Pacific are a subset of the genetic diversity found in Eurasia which itself is a subset of the diversity found in Africa.

It is clear that the human species is relatively young. As a species, we most likely originated in east Africa, according to most archaeologists, 100,000 to 300,000 years ago, and diverged as groups, expanded, moved, and settled the globe. During these migrations, and in the time since, there has been some degree of independent evolution of the populations that settled the various continents of the world. The simplest evidence of this evolution can be seen in the differences in allele frequencies at genetic markers. Generally, we see that alleles found in one population are also found in all populations and the alleles that are the most common in one are also common in others. These similarities between populations highlight the recent common origin of all populations and strong connections between populations throughout human history. However, there are examples of genetic markers which are different between populations and it is these markers, called Ancestry Informative Markers (AIMs), which can be used to estimate the ancestral origins of a person or population. Race is a complex and multivariate construct that we tend to over simplify but in our analysis, we are measuring a person's genetic ancestry and not their race. Your DNA has no recorded history of your political, social, personal or religious beliefs. It is a simple four letter code that records all of the changes in the DNA from one generation to the next. We report those changes, they are like finger prints and snow flakes, unique and wildly complex.

14. Michèle Lamont has described the emotional pragmatism with which academic peer review panels approach the task of judging academic excellence. Lamont, *How Professors Think: Inside the Curious World of Academic Judgment* (Cambridge, MA: Harvard University Press, 2010).

15. *Nature Genetics*, "Census, Race, and Science," 24 (2000), 97–98.

16. Chiara Romualdi, David Balding, Ivane S. Nasidze, Gregory Risch, et al., "Patterns of Human Diversity, Within and Among Continents, Inferred from Biallelic DNA Polymorphisms," *Genome Research* 12 (2002), 602–12.

17. James F. Wilson, Michael E. Weale, Alice C. Smith, Fiona Gratrix, et al., "Population Genetic Structure of Variable Drug Response," *Nature Genetics* 29 (2001), 265–69. Though these scientists used the word "ethnicity," they were looking for vast continental differences in their clusters. Thus it is more appropriate to say they were examining the genome for evidence of race.

18. *Ibid.*, 266.

19. Howard McLeod, "Pharmacogenetics: More Than Skin Deep," *Nature Genetics* 29 (2001), 248.

20. *Nature Genetics*, "Genes, Drugs and Race," 29 (2001), 240.

21. The *Journal of Adolescent Health* advised contributors to carefully consider the rationale and consequences of using race, while the *Archives of Pediatrics and Adolescent Medicine* issued a policy barring the use of race and ethnicity as explanatory variables. Iris F. Litt, "When Race Matters," *Journal of Adolescent Health* 29 (2001), 311; Frederick P. Rivara and Laurence Finberg, "Use of the Terms Race and Ethnicity," *Archives of Pediatrics and Adolescent Medicine* 155 (2001), 119. *Genomics* celebrated a recent publication by evolutionary scientist Joseph Graves Jr. warning researchers of "genetically 'locking in' a group of people." Fintan R. Steele, "Genetic 'Differences,'" *Genomics* 79 (2002), 145.

22. Neil Risch, Esteban Burchard, Elad Ziv, and Hua Tang, "Categorization of Humans in Biomedical Research: Genes, Race and Disease," *Genome Biology* 3 (2002), 1–12.

23. W. Kalow, "Both Populations and Individuals Are Evolutionary Targets: Pharmacogenomic and Cultural Indicators," *Pharmacogenomics Journal* 2 (2002), 12–14. Kalow argued that both populations and individuals were relevant targets.

24. Noah A. Rosenberg, Jonathan K. Pritchard, James L. Weber, Howard M. Cann, et al., "Genetic Structure of Human Populations," *Science* 298 (2002), 2381–85.

25. The *New York Times* also reported a study claiming that Franz Boas misinterpreted the environmental effect on the bodies of recent immigrants. The study found there were significant morphological differences in skulls that followed continental origins. Nicholas Wade, "A New Look at Old Data May Discredit a Theory on Race," *New York Times* (Oct. 8, 2002); *Toronto Star*, "Race Seen as Crucial to Medical Research" (Aug. 2, 2002), F5; *Los Angeles Times*, "Genetics Study Identifies Five Ancestry Groups" (Dec. 20, 2002), 6C.

26. Nicholas Wade, "The Palette of Humankind," *New York Times* (Dec. 24, 2002), 3.

27. Nicholas Wade, "Race Is Seen as Real Guide to Track Roots of Disease."

28. *Ibid.*

29. Pfaff, "Race Reemerges in a PC World," A11. Also see Nicholas Wade, "Genetics Study Identifies Five Main Populations," *New York Times* (Dec. 20, 2002), 37. Articles like this ran in the *International Herald Tribune, Los Angeles Times,* and *Boston Globe.*

30. *Los Angeles Times,* "Genetics Study Identifies Five Ancestry Groups."

31. *Ibid.* Stanford's bioethicist Mildred Cho was quoted in the *Los Angeles Times* as opposing the constant search for a biological basis for race. "The core question is why is it we keep looking at genes and variations and organizing them into categories?" Alan Goodman stated that the clusters were driven by geography. Also see American Sociological Association, *The Importance of Collecting Data and Doing Social Scientific Research on Race* (Washington, DC: American Sociological Association 2003).

32. Wade, "Race Is Seen as Real Guide to Track Roots of Disease."

33. *Ibid.*

34. Rick Weiss, "Genome Project Completed; Findings May Alter Humanity's Sense of Itself, Experts Predict," *Washington Post* (Apr. 15, 2003), A06.

35. *Ibid.* Also see Nell Boyce, "All the Difference in the World," *U.S. News* (Nov. 3, 2002).

36. The long-awaited results of the VaxGen HIV vaccine trial suggested that the vaccine worked only in African Americans, while BiDil launched its end-stage clinical trial to be the first race-based medicine. Though end-stage clinical trials do not employ genomic methods, they have a direct effect on how future genomic inquiries are framed. Their reports also suggest to the public that there is a biological basis for racial health disparities. Finally, in ordering therapeutic intervention by race, they stratify subject experiences by race—this *produces* a biological racial effect in society.

37. CBS News, "Genetic Double-Whammy for Blacks" (Oct. 10, 2002); Susan Brink, "An Unhealthy Combo," *U.S. News* (Mar. 3, 2002). CBS News also reported that lactose intolerance was in genes possessed by "people of all racial groups." From prostate cancer and hypertension to heart failure and smoking addiction—the article claimed African American disparities were due to genetics. CBS Morning News, "Researchers Say Lactose Intolerance Is All in the Genes" (Jan. 14, 2002).

38. Elad Ziv and Esteban Gonzalez Burchard, "Human Population Structure and Genetic Association Studies," 432.

> First, in general, genetic differentiation among indigenous groups follows geographical distance. The greatest differentiation follows continent of ancestral origin. Thus, classification of individuals into major racial classifications does provide information about genetic ancestry. Second, there is, to varying degrees population substructure within major groups. In general, the substructure of human populations within the major groups also tended to follow ethnic lines although some exceptions were found among closely related populations, such as Europe and the Middle East. Finally, in populations which have migrated and have undergone admixture especially in the US and Latin America and along the border regions of continents, self-described ancestry may be a less accurate predictor of genetic ancestry.

39. Heather E. Collins-Schramm, Bill Chima, Darwin J. Operario, Lindsey A.

Criswell, et al., "Markers Informative for Ancestry Demonstrate Consistent Megabase-Length Linkage Disequilibrium in the African American Population," *Human Genetics* 113 (2003), 211–19; Noah A. Rosenberg, Lei M. Li, Ryk Ward, and Jonathan K. Pritchard, "Informativeness of Genetic Markers for Inference of Ancestry," *American Journal of Human Genetics* 73 (2003), 1402–22.

40. Sally Satel, "I Am a Racially Profiling Doctor," *New York Times* (May 5, 2002); "One Nation Under Racist Doctors? Don't Believe the Media Hype," *Wall Street Journal* (Apr. 4, 2002). Satel's "hype" article argued that it was unclear whether there were disparities in care and conjectured that blacks might simply opt out of treatment for unexplained reasons. It claimed that false stories made blacks mistrustful of medicine.

41. Satel, "I Am a Racially Profiling Doctor." This article is included in *The Best American Science Writing 2002*. Satel has also gone on to battle the idea of the existence of racial health disparities. See her website for a list of books and articles: http://www.sallysatelmd.com.

42. Andrew Smart, Richard Tutton, Richard Ashcroft, Paul Martin, et al., "Social Inclusivity vs Analytical Acuity? A Qualitative Study of UK Researchers Regarding the Inclusion of Minority Ethnic Groups in Biobanks," *Medical Law International* 9 (2008), 169–90.

43. Elizabeth G. Phimister, "Medicine and the Racial Divide," *New England Journal of Medicine* 348 (2003), 1082.

44. Wade, "Race Is Seen as Real Guide to Track Roots of Disease."

45. Esteban González Burchard, Elad Ziv, Natasha Coyle, Scarlett Lin Gomez, et al., "The Importance of Race and Ethnic Background in Biomedical Research and Clinical Practice," *New England Journal of Medicine* 348 (2003), 1172.

46. *Ibid.*, 1174.

47. Richard S. Cooper, Jay S. Kaufman, and Ryk Ward, "Race and Genomics," *New England Journal of Medicine* 348 (2003), 1166–70.

48. *Ibid.*, 1167.

49. This dispute continued in the *International Journal of Epidemiology*, where Cooper appealed to basic researchers to keep their work separate from public health. Karter contrarily suggested that basic researchers start with racial analysis and simply pool samples if no race-based variation was found. Andrew J. Karter, "Race, Genetics, and Disease—In Search of a Middle Ground," *International Journal of Epidemiology* 32 (2003), 27. In a subsequent article in *Diabetes Care*, Karter hailed the successful legacy of using race in epidemiology. He briefly introduced a point that has gotten more attention of late: the idea of a social-biological feedback loop due to racism in the wider society. Karter, "Race and Ethnicity: Vital Constructs for Diabetes Research," *Diabetes Care* 26 (2003), 2191.

50. Cooper, Kaufman, and Ward, "Race and Genomics," 1169. In fact, James Watson was again criticized in the media for making more incendiary comments. This time he argued for intelligence and beauty-driven eugenics: "If you really are stupid, I would call that a disease. . . . People say it would be terrible if we made all girls pretty. I think it would be great." Nobel Laureate and cofounder of the Human Genome Project, Sir John

Sulston, warned that Watson was in dangerous territory but defended the scientist's right to speak on ethical quandaries. *Agence France Press,* "Manipulate Genes to Boost Intelligence, Beauty; DNA's Co-discoverer" (Feb. 28, 2003); Mark Henderson, "Let's Cure Stupidity, Says DNA Pioneer," *Times* (Feb. 28, 2003), 13.

51. As the debate reverberated in the media, the field's uncertainty about race came through in contrasting reports from various news desks. The *New York Times* explained that diseases "may be caused by different genes in different racial groups" and cited Risch as hoping that. Contrarily, in the *Guardian* (U.K.), genetic immunologist Johnjoe McFadden wrote an op-ed depicting Cavalli-Sforza's work and the work of other recent genomicists as proving race insignificant. Similarly, in a celebratory article summing "23 Ways That DNA Changed the World," the *Independent* put a final end to the "myth of race" at number twelve. Nicholas Wade, "Two Scholarly Articles Diverge on Role of Race in Medicine," *New York Times* (Mar. 20, 2003); Johnjoe McFadden, "Written in Our Genes," *Guardian* (Mar. 31, 2003); Steve Conner, "23 Ways That DNA Changed the World," *Independent* (Feb. 27, 2003), 4.

52. *Nature Genetics,* "The Unexamined 'Caucasian,'" 36 (2004), 541.

53. Margaret Winker, "Measuring Race and Ethnicity: Why and How?" *Journal of the American Medical Association* 292 (2004), 1614.

54. Nicholas Wade, "Articles Highlight Different Views on Genetic Basis of Race," *New York Times* (Oct. 29, 2004).

55. Sharon Begley, "Ancestry Trumps Race in Predicting Efficacy of Drug Treatments," *Wall Street Journal* (Oct. 29, 2004), B1.

56. Roger Highfield, "'Ethnic Drug' Raises Fears over Race and Genetics," *Daily Telegraph* (Nov. 1, 2004), 13.

57. *New York Times,* "Toward the First Racial Medicine" (Nov. 13, 2004), 14.

58. Cohn said, "I don't pretend to understand all the factors. And I don't suggest it's a uniform difference. But on average, responses appear to be different." David Rotman, "Race and Medicine," *Technology Review* (Apr. 1, 2005).

59. William E. Evans and Mary V. Relling, "Moving Towards Individualized Medicine with Pharmacogenomics," *Nature* 429 (2004), 467.

60. Robin Marantz Henig, "The Genome in Black and White (and Gray)," *New York Times* (Oct. 10, 2004), 47; *New York Times,* "Toward the First Racial Medicine" (Nov. 13, 2004), 14; Nicholas Wade, "Race-Based Medicine Continued..." *New York Times* (Nov. 14, 2004), 12.

61. Wade, "Race-Based Medicine Continued..." 12.

62. Henig, "Genome in Black and White (and Gray)."

63. *Ibid.*

64. Sally Lehrman, "The Reality of Race," *Scientific American* (Jan. 1, 2003). On the problems with early forensic DNA sequencing and the lingering use of race despite forensic DNA advances, see Jonathan Kahn, "What's the Use of Race in Presenting Forensic DNA Evidence in Court?" in *What's the Use of Race? Modern Governance and the Biology of Difference,* edited by Ian Whitmarsh and David S. Jones (Cambridge, MA: MIT Press, 2010).

65. Troy Duster, "Buried Alive! The Concept of Race in Science," in *Genetic Nature/ Culture: Anthropology and Science Beyond the Two-Culture Divide*, edited by Alan H. Goodman, Deborah Heath, and M. Susan Lindee (Berkeley: University of California Press, 2003), 258–77; "The Sociology of Science and the Revolution in Molecular Biology," in *The Blackwell Companion to Sociology*, edited by Judith R. Blau (New York: Blackwell, 2001). Troy Duster and Dorothy Nelkin presented their views on the social implications of genomics at the Consortium of Social Science Associations' "The Genetic Revolution and the Meaning of Life: How Will Society Respond to the Explosion of Knowledge?" Past president of the American Anthropological Association and the American Association of Physical Anthropologists, George Armelagos, encouraged genomicists to drop the issue of race once and for all. Ryan A. Brown and George J. Armelagos, "Apportionment of Racial Diversity: A Review," *Evolutionary Anthropology* 10 (2001), 34–40. Alan Goodman, a future president of the American Anthropological Association likewise argued that the genetic data did not support the idea of race, thus attributing any postgenomic persisting belief in race to ideologies of genetic essentialism and scientific racialism. Alan Goodman, "Why Genes Don't Count (for Racial Differences in Health)," *American Journal of Public Health* 90 (2000), 1699–1702. Goodman organized a parallel conference to one of the landmark Human Genome Diversity Project North American conferences in efforts to establish a scientific alternative.

66. Carl Elliot and Paul Brodwin, "Identity and Genetic Ancestry Tracing," *British Medical Journal* 325 (2002), 1469–71; Josephine Johnston, "Resisting a Genetic Identity: The Black Seminoles and Genetic Tests of Ancestry," *Medicine and Ethics* 3 (2003), 262–71.

67. Josephine Johnston and Mark Thomas, "The Science of Genealogy by Genetics," *Developing World Bioethics* 3 (2003), 103–8.

68. George T.H. Ellison and Ian Rees Jones, "Social Identities and the 'New Genetics': Scientific and Social Consequences," *Critical Public Health* 12 (2002), 266.

69. *Ibid.*, 255.

70. Celeste Michelle Condit, Roxanne Parrott, and Tina M. Harris, "Lay Understandings of the Relationship Between Race and Genetics: Development of a Collectivized Knowledge Through Shared Discourse," *Public Understanding of Science* 11 (2002), 373–87.

71. Celeste Condit, Alan Templeton, Benjamin R. Bates, Jennifer L. Bevan, et al., "Attitudinal Barriers to Delivery of Race-Targeted Pharmacogenomics Among Informed Lay Persons," *Genetics in Medicine* 5 (2003), 385–92; Benjamin R. Bates, Alan Templeton, P. J. Achter, Tina M. Harris, et al., "What Does 'A Gene for Heart Disease' Mean?" *American Journal of Medical Genetics* 119A (2003), 156–61; Roxanne L. Parrott, Kami J. Silk, and Celeste Condit, "Diversity in Lay Perceptions of the Sources of Human Traits: Genes, Environments, and Personal Behaviors," *Social Science and Medicine* 56 (2003), 1099–109.

72. The gist of the critique centered on sociological questions about colorblind science and the potential reification of race in an uncritically racialist science. David Skinner and Paul Rosen called it "opening up the white box" of science. Skinner and Rosen, "Opening the White Box: The Politics of Racialised Science and Technology," *Science as*

Culture 10 (2001), 285–300. Also see Duster, "The Sociology of Science and the Revolution in Molecular Biology" and "Buried Alive!"; Goodman, "Why Genes Don't Count."

73. Goodman, "Why Genes Don't Count"; Brown and Armelagos, "Apportionment of Racial Diversity."

74. Lundy Braun, "Race, Ethnicity, Health: Can Genetics Explain Disparities?" *Perspectives in Biology and Medicine* 45 (2002), 159–74; Jacqueline Stevens, "Racial Meanings and Scientific Methods: Changing Policies for NIH-Sponsored Publications Reporting Human Variation," *Journal of Health Politics, Policy and Law* 28 (2003), 1037–87;

"Symbolic Matter: DNA and Other Linguistic Stuff," *Social Text* 20 (2003), 105–36; Gannett, "Racism and Human Genome Diversity Research."

75. Michael Root, "The Problem of Race in Medicine," *Philosophy of the Social Sciences* 31 (2001), 21; "The Use of Race in Medicine as a Proxy for Genetic Differences," *Philosophy of Science* 70 (2003), 1176–77.

76. Stevens, "Racial Meanings and Scientific Methods," 1035.

77. Richard S. Cooper, "Race, Genes, and Health: New Wine in Old Bottles?" *International Journal of Epidemiology* 32 (2003), 23–25.

78. Reardon, "Decoding Race and Human Difference in a Genomic Age," 38.

79. Gannett, "Racism and Human Genome Diversity Research," S479.

80. Also see Lisa Gannett, "Group Categories in Pharmacogenetics Research," *Philosophy of Science* 72 (2005), 1232–47; Jenny Reardon, "Genomics and Race: Breaking Barriers or Propagating Racism?" *Ethos* (2004), 6–11.

81. Anne Fausto-Sterling, "Refashioning Race: DNA and the Politics of Health Care," *differences* 15 (2004), 1–37; Braun, "Race, Ethnicity, Health: Can Genetics Explain Disparities?"; Wald, "Race- Based Genomics Research."

82. Fausto-Sterling, "Refashioning Race," 28.

83. Braun, "Race, Ethnicity, Health," 170.

84. Cf. Joan H. Fujimura, Troy Duster, and Ramya Rajagopalan, "Matters of Consequence—Race, Genetics, and Disease: Questions of Evidence," *Social Studies of Science* 38 (2008), 650.

85. Noah A. Rosenberg, Saurabh Mahajan, Sohini Ramachandran, Chengfeng Zhao, et al., "Clines, Clusters, and the Effect of Study Design on the Inference of Human Population Structure," *PLoS Genetics* 1 (2005), e70.

86. See the conclusion of this chapter for a discussion of subsequent strands of resistance to the sociogenomic consensus.

87. Robin O. Andreasen, "The Cladistic Race Concept: A Defense," *Biology and Philosophy* 19 (2004), 442.

88. Claudia Dreifus, "A Sociologist Confronts 'The Messy Stuff,'" *New York Times* (Oct. 18, 2005).

89. Troy Duster, "Feedback Loops in the Politics of Knowledge Production," *Theorie und Praxis* 13 (2004), 45.

90. Flora A. Ukoli, Jay H. Fowke, Phillip Akumabor, Temple Oguike, et al., "The Association of Plasma Fatty Acids with Prostate Cancer Risk in African Americans and Africans," *Journal of Health Care for the Poor and Underserved* 21 (2010), 127–47.

91. As Chakravarti put it, "In some sense I'm forced to use the word ["race"], because there is no other word more often that I use."

92. Amy Harmon, "In DNA Era, New Worries About Prejudice," *New York Times* (Nov. 11, 2007).

93. Jonathan Michael Kaplan, "When Socially Determined Categories Make Biological Realities: Understanding Black/white Health Disparities in the U.S." *The Monist* 93 (2010), 283–99.

94. Pamela Sankar, Mildred K. Cho, Celeste M. Condit, Linda M. Hunt, et al., "Genetic Research and Health Disparities," *Journal of the American Medical Association* 291 (2004), 2985–89.

95. See, for example, Nancy Krieger, Jarvis T. Chen, and Pamela D. Waterman, "Temporal Trends in the Black/White Breast Cancer Case Ratio for Estrogen Receptor Status: Disparities Are Historically Contingent, not Innate," *Cancer Causes Control* 22 (2010), 511–14.

96. Anne Fausto-Sterling, "Nature VERSUS Nurture (Part 1): It's Time to Withdraw from This War!" *Sexing the Body* (blog) *Psychology Today* (July 29, 2010), http://www.psychologytoday.com/blog/sexing-the-body/201007/nature-versus-nurture-part-1-it-s-time-withdraw-war; Troy Duster, "Welcome, Freshmen. DNA Swabs, Please," *Genetic Watchdog* (blog), *Council for Responsible Genetics* (May 30, 2010), http://www.councilforresponsiblegenetics.org/blog/post/Welcome-Freshmen-DNA-Swabs-Please.aspx.

97. Shelley L. Berger, Tony Kouzarides, Ramin Shiekhattar, and Ali Alishilatifard, "An Operational Definition of Epigenetics," *Genes and Development* 23 (2009), 781–83.

98. On narrating the past with present concerns, see Danto, *Narration and Knowledge*, 143–200. On self-accounts and coherence, see Joseph C. Hermanowicz, *The Stars Are Not Enough: Scientists: Their Passions and Professions* (Chicago: University of Chicago Press, 1998).

99. David Goldstein, *Jacob's Legacy: A Genetic View of Jewish History* (New Haven, CT: Yale University Press, 2009).

100. In addition to his appearances in articles about new biomedical advances, the *New York Times* interviewed Goldstein for its "Scientists at Work" series. Nicholas Wade, "A Dissenting Voice as the Genome Is Sifted to Fight Disease," *New York Times* (Sept. 15, 2008), F3.

101. Kelly Moore, *Disrupting Science* (Princeton, NJ: Princeton University Press, 2008), 193.

102. *Ibid.*, 198.

103. Moore explains how different generations struggled with the definition of responsibility and drew on concurrent activist tactics to operationalize it. For example, from the 1950s to the 1970s, nuclear scientists took on more direct forms of science activism, from bearing witness to problems with nuclear technology and publicizing its social costs to engaging in protests.

104. *Ibid.*, 10.

105. Scott Frickel, *Chemical Consequences: Environmental Mutagens, Scientist Activism, and the Rise of Genetic Toxicology* (New Brunswick, NJ: Rutgers University Press,

2004); Edward Woodhouse and Steve Breyman, "Green Chemistry as Social Movement?" *Science, Technology, and Human Values* 30 (2005), 199–222.

106. Omi and Winant, *Racial Formations in the United States*, 55–56.

107. Erving Goffman, *The Presentation of Self in Everyday Life* (Gloucester, MA: Peter Smith, 1999).

108. See Richard C. Lewontin, Steven P.R. Rose, and Leon J. Kamin, *Not in Our Genes* (New York: Pantheon Books, 1984); Stephen Jay Gould, *The Mismeasure of Man* (New York: Norton, 1996 [1981]).

109. Reardon, *Race to the Finish*.

110. Scientists often make conceptual distinctions between "biology" and "genetics," whereby biology refers to phenotype (morphology, physiology, and possibly behavior) and genetics refers to genetic or genomic structure and function.

111. Ian Whitmarsh develops the notion of biomedical oscillations in his work on Caribbean racial genomics. See "Hyperdiagnostics: Postcolonial Utopics of Race-Based Biomedicine," *Medical Anthropology* 28:3 (2009), 285–315. I explore the three dominant working models of race in the next chapter.

112. In *Objectivity* (New York: Zone Books, 2007), Lorraine Daston and Peter Galison distinguish between various antisubjective scientific styles of practice: truth-to-nature, mechanical objectivity, and trained judgment. The style of practice I describe complicates the very distinction between objectivity and subjectivity that these models presuppose.

113. Robert Merton, *The Sociology of Science: Theoretical and Empirical Investigations* (Chicago: University of Chicago Press, 1979 [1942]).

114. Charles Rotimi, "Genetic Ancestry Tracing and the African Identity: A Double-Edged Sword?" *Developing World Bioethics* 3 (2003), 151–58.

115. Rebecca L. Lamason, Manzoor-Ali P.K. Mohideen, Jason R. Mest, Andrew C. Wong, et al., "SLC24A5, a Putative Cation Exchanger, Affects Pigmentation in Zebrafish and Humans," *Science* 310 (2005), 1782–86.

116. D. Gerasimos, I. Stefanaki, V. Nikolaou, M. Poulou, et al., "Mutations of the Golden Gene in a Melanoma Case/Control Study," *Melanoma Research* (2006), S23–S24; Jessica L. Moore, Lindsay M. Rush, Carol Breneman, Manzoor-Ali P.K. Mohideen, et al., "Zebrafish Genomic Instability Mutants and Cancer Susceptibility," *Genetics* 174 (2006), 585–600; Brian McEvoy, Sandra Beleza, and Mark D. Shriver, "The Genetic Architecture of Normal Variation in Human Pigmentation: An Evolutionary Perspective and Model," *Human Molecular Genetics* 15 (2006), R176–R181; Keith C. Cheng and Victor A. Canfield, "The Role of SLC24A5 in Skin Color," *Experimental Dermatology* 15 (2006), 836–38. In fact, pigmentation has since become a very important topic to the field. Perlegen and deCODE have each run genome-wide association studies to find variants responsible for skin, eye, and hair color. Renee P. Stokowski, P.V. Krishna Pant, Tony Dadd, Amelia Fereday, et al., "A Genomewide Association Study of Skin Pigmentation in a South Asian Population," *American Journal of Human Genetics* 81 (2007), 1119–32; Patrick Sulem, Daniel F. Gudbjartsson, Simon N. Stacey, Agnar Helgason, Thorunn Rafnar, et al., "Genetic Determinants of Hair, Eye and Skin Pigmentation in Europeans," *Nature Genetics* 39 (2007), 1443–52. A prominent dermatology journal has also presented its first issue

considering the genomics of pigmentary disorders (see *Dermatologic Clinics*, July 2007), and a letter to the *Journal of the American Academy of Dermotology* has raised the issue of abandoning race for genomic reasons. Nonhlanhla P. Khumalo, "Yes, Let's Abandon Race—It Does Not Accurately Correlate with Hair Form," *Journal of the American Academy of Dermatology* 56 (2007), 709–10. Many laboratories have researched the relationship between several pigmentation genes and continental variation. Sean Myles, Mehmet Somel, Kun Tang, Janet Kelso, et al., "Identifying Genes Underlying Skin Pigmentation Differences Among Human Populations," *Human Genetics* 120 (2007), 613–21; O. Lao, J.M. de Gruijter, K. van Duijn, A. Navarro, et al., "Signatures of Positive Selection in Genes Associated with Human Skin Pigmentation as Revealed from Analyses of Single Nucleotide Polymorphisms," *Annals of Human Genetics* 71 (2007), 354–69. Two Kurume University geneticists have offered an ancestry informative marker panel based on the "zebrafish" gene. Mikiko Soejima and Yoshiro Koda, "Population Differences of Two Coding SNPs in Pigmentation-Related Genes SLC24A5 and SLC45A2," *International Journal of Legal Medicine* 121 (2007), 36–39. It remains to be seen how this evidence of pigmentation variation along continental lines will affect future terms of the debate. It's important to note that the structure of all studies predicated on haplotypic data (Hap-Map or Perlegen samples) are constrained to use a continental-comparative scheme that references global diversity in terms of Western Europe, Far East Asia, and West Africa.

117. Cheng and Canfield, "Role of SLC24A5 in Skin Color," 837.

Chapter 4

1. I adopt the notion of "working models" from Ann Morning, *The Nature of Race: How Scientists Think and Teach About Human Difference* (Berkeley: University of California Press, 2011). Also see, "Toward a Sociology of Racial Conceptualization for the 21st Century," *Social Forces* 87 (2009), 1167–92.

2. As defined in the Introduction, "typological" refers to a racial classification system that conceives of races as different human types.

3. Robert Bernasconi and Tommy Lott, *The Idea of Race* (Indianapolis, IN: Hackett, 2000).

4. Steven Jay Gould, *Mismeasure of Man*.

5. Charles Darwin, *On the Origin of Species by Means of Natural Selection, or the Preservation of Favoured Races in the Struggle for Life* (London: John Murray, 1859); *The Descent of Man, and Selection in Relation to Sex* (London: John Murray, 1871). The Darwinian population concept is the primary zoological concept held for animal species.

6. Reardon, *Race to the Finish*, 17–44. The movement to bring Darwinian evolutionism in line with genetics is called "the modern evolutionary synthesis." Scientists who led the synthesis were active participants in the UNESCO forums that produced the 1950 and 1952 Statements on Race.

7. Lamont, *How Professors Think*.

8. For a discussion of the role of racial memory and practice in everyday perception, see Sarah Daynes and Orville Lee, *Desire for Race* (Cambridge, UK: Cambridge University Press, 2008).

9. Morning, "Toward a Sociology of Racial Conceptualization for the 21st Century," 1168.

10. See, for example, Susan Leigh Star and Geoffrey Bowker, *Sorting Things Out* (Cambridge, MA: MIT Press, 1999).

11. In animal conservation, scientists combine characteristics like tail shape and coat pattern with genomic traits to formulate taxonomies around which one can then politically organize.

12. Smith was one of two foremost developers of admixture mapping technology at the time of publication of the human genome. Today the ancestry informative marker panels he developed are used in labs across the world on diseases as wide-ranging as congenital heart failure, prostate cancer, and alcoholism. Note the subspecies view that admixture engenders: according to this model, African American population ancestry does not possess enough indigenous American DNA to merit its analysis in individuals. This is a problematic assumption, since different historical populations have had varying experiences cohabiting, creating families, and engaging in exogamy.

13. Daynes and Lee, *Desire for Race.*

14. I did not ask scientists to provide a racial or ethnic identification; therefore, the following figures for identification represent the proportion of scientists who asserted an identification without prompt.

15. Ruha Benjamin discusses "by the people, for the people" tropes in genomics as a dominant form of bio-constitutionalism in *People's Science* (Stanford, CA: Stanford University Press, forthcoming).

16. Kwame Anthony Appiah, *The Ethics of Identity* (Princeton, NJ: Princeton University Press, 2005).

17. On epidemiological rationales for the use of race in cardiovascular disease, see Janet K. Shim, "Constructing 'Race' Across the Science-Lay Divide: Racial Formation in the Epidemiology and Experience of Cardiovascular Disease," *Social Studies of Science* 35 (2005), 405–36.

18. Burchard, Ziv, Coyle, Gomez, et al., "Importance of Race and Ethnic Background in Biomedical Research and Clinical Practice."

19. Morning shows biological and cultural essentialisms working in tandem in college student articulations of racial beliefs. Morning, *The Nature of Race.*

20. Geneticist A.W.F. Edwards, biostatistician Laurent Excoffier, and philosopher of statistics Ian Hacking have popularized this counterargument to Lewontin's claim. See A.W.F. Edwards, "Human Genetic Diversity: Lewontin's Fallacy," *BioEssays* 25 (2003), 801; Laurent Excoffier, "Human Diversity: Our Genes Tell Where We Live," *Current Biology* 13 (2003), R135; and Ian Hacking, "Genetics, Biosocial Groups and the Future of Identity," *Daedalus* (Fall 2006), 81–95.

21. Without prompting, about half of the genomicists I interviewed told me they were Jewish. Like the scientists who said they were part of multiracial or minority families, talking about their personal racial identities and motivations seemed to bring them into a more reflective and thoughtful mind frame to talk about their work. These scientists also gave the impression that they were ethically confident that their research was

publicly sensitive and relevant. During the course of our conversations, they traveled between genetic considerations and concerns about the troubled history of eugenics. They attempted to carve the "realities" from the "lies," explaining how they saw biology moving beyond racist genetics.

22. A cline is a gradual change in a trait across the range of variation in a species or population.

23. See his 2009 essay: Aravinda Chakravarti, "Being Human: Kinship: Race Relations," *Nature* 457 (2009), 380–81.

Chapter 5

1. Epstein, *Inclusion*.

2. Paraphrased in *Science Daily*. European Society of Human Genetics, "Understanding Genetic Mixing Through Migration: A Tool for Clinicians as Well as Geneaologists," *Science Daily* (June 21, 2010), http://www.sciencedaily.com/releases/2010/06/100611204144 .htm.

3. Kenneth J. Cooper, "Stanford Geneticist Pushes for More African-Americans, Hispanics to Join Critical Research," *Black Voice News* (Apr. 25, 2011), http://www.black-voicenews.com/news/news-wire/46057-stanford-geneticist-pushes-for-more-african-americans-hispanics-to-join-critical-research.html.

4. *Ibid.*

5. Of Kittles, Bustamante has claimed, "To my knowledge, he's the only African-American geneticist in the country who has any credibility" (*Ibid.*).

6. For studies of sampling practices trained on nonelites, see M'Charek, *Human Genome Diversity Project*; Fullwiley, "Race and Genetics" and "Molecularization of Race"; Hunt and Megyesi, "Ambiguous Meanings of the Racial/Ethnic Categories Routinely Used in Human Genetics Research."

7. Pierre Bourdieu, *Masculine Domination* (Stanford, CA: Stanford University Press, 1997), 97.

8. The Genetic Investigation of Anthropometric Traits [GIANT] Consortium draws on pooled data from over 180,000 samples. For details, see http://childrenshospital.org/ newsroom/Site1339/mainpageS1339P663.html.

9. Daynes and Lee, *Desire for Race*.

10. *Ibid.*

11. The project is called The Cancer Genome Atlas, or TCGA.

12. George Ayodo, Alkes L. Price, Alon Keinan, Arthur Ajwang, et al., "Combining Evidence of Natural Selection with Association Analysis Increases Power to Detect Malaria-Resistance Variants," *American Journal of Human Genetics* 81 (2007), 234–42.

13. This assumption can be questioned, since migrations have occurred throughout human history. See, for example, Tenzin Gayden, Alicia M. Cadenas, Maria Regueiro, Nanda B. Singh, et al., "The Himalayas as a Directional Barrier to Gene Flow," *American Journal of Human Genetics* 80 (2007), 884–94.

14. The Multiethnic Cohort, a large prospective cohort that was founded in the mid 1990s, is produced by a partnership between the University of Southern California, the

University of Hawai'i, the Massachusetts Institute of Technology, and Harvard Medical School. The study offers research subjects the choice to identify as African American, Japanese American, Native Hawaiian, Latino, or European American. See L. N. Kolonel, B. E. Henderson, J. H. Hankin, A. M. Nomura, et al., "A Multiethnic Cohort in Hawaii and Los Angeles: Baseline Characteristics," *American Journal of Epidemiology* 151 (2000), 346–57.

15. On Burchard's ongoing work with the GALA study, see http://bts.ucsf.edu/burchard. On Mountain's former collections, see http://www.stanford.edu/group/mountainlab/people/joanna_mountain.html.

16. Researchers used to collect DNA that could be immortalized through recombinant DNA procedures and storage in a biobank. Now those samples and many others collected by large-scale population studies are made available in sufficient amounts by national biobanks.

17. Reardon, *Race to the Finish.*

18. For a list of projects, see http://crggh.nih.gov/projects.cfm.

19. Cultural anthropologists alternatively define "salvage anthropology" as rushing to do research before a particular culture dies out or becomes "modernized," and losing all the ritual practices in the process.

20. In *Race to the Finish*, Reardon presents the divisions between molecular and cultural scientists within anthropology and their opposing views about how to work with minority communities.

21. Appiah, *Ethics of Identity*, 21.

22. On trust in the context of genomic studies at the U.S.-Mexico border, see Montoya, *Making the Mexican Diabetic*, 78.

23. Nila Patil, Anthony J. Berno, David A. Hinds, Wade A. Barett, et al., "Blocks of Limited Haplotype Diversity Revealed by High Resolution Scanning of Human Chromosome 21," *Science* 294 (2001), 1719–23.

24. David A. Hinds, Laura L. Stuve, Geoffrey B. Nilsen, Eran Halperin, et al., "Whole-Genome Patterns of Common DNA Variation in Three Human Populations," *Science* 307 (2005), 1072–79.

25. See Gates, *African American Lives* (miniseries).

26. See his lab's list of projects at http://depts.washington.edu/bamshad/research/evolution.html.

27. Burchard, Ziv, Coyle, Gomez, et al., "Importance of Race and Ethnic Background in Biomedical Research and Clinical Practice."

28. Teri A. Manolio, Joan E. Bailey-Wilson, and Francis S. Collins, "Genes, Environment and the Value of Prospective Cohort Studies," *Nature Reviews Genetics* 7 (2005), 812–20.

29. Also see Fujimura and Rajagopalan, "Different Differences."

30. A helpful table of the research process from sampling to therapy is provided in Montoya, *Making the Mexican Diabetic*, 46.

31. My claim is based on content analysis of scientists' publications from 1986 to 2010 and preliminary findings from a pilot study Aaron Panofsky and I have conducted on taxonomic strata in genomic publications.

32. A. L. Price, J. Butler, N. Patterson, C. Capelli, et al., "Discerning the Ancestry of European Americans in Genetic Association Studies," *PLoS Genetics* 4 (2008), e236.

33. For a review, see Andrew Smart, Richard Tutton, Paul Martin, George T.H. Ellison, and Richard Ashcroft, "The Standardization of Race and Ethnicity in Biomedical Biomedicine Editorials and UK Biobanks," *Social Studies of Biomedicine* 38 (2008), 407–23.

34. In 2008 the *Journal of the American Medical Association*'s deputy editor, Margaret Winker, presented data at the MIT "What's the Use of Race?" conference, suggesting that the journal's 2004 policy had not been fruitful.

35. Pamela Sankar, Mildred K. Cho, and Joanna Mountain, "Race and Ethnicity in Genetic Research," *American Journal of Medical Genetics* 143 (2007), 961–70.

36. For a discussion of tradeoffs between social inclusivity and analytical acuity, see Andrew Smart, Richard Tutton, Richard Ashcroft, Paul Martin, et al., "Social Inclusivity vs. Analytical Acuity? A Qualitative Study of UK Researchers Regarding the Inclusion of Minority Ethnic Groups in Biobanks," *Medical Law International* 9 (2008), 169–90.

37. Duana Fullwiley's forthcoming book on American genomics will provide the first interlab comparisons.

38. Guido Barbujani and Vincenza Colonna, "Human Genome Diversity: Frequently Asked Questions," *Trends in Genetics* 26 (2010), 285–95.

39. Pierre Bourdieu, *Pascalian Meditiations* (Stanford, CA: Stanford University Press, 1998), 13.

40. Epstein, *Inclusion*, 142.

41. *Ibid.*, 143.

Chapter 6

1. David J. Hess, *Science Studies: An Advanced Introduction* (New York: New York University Press, 1997).

2. Gary Lee Downey and Joseph Dumit, *Cyborgs and Citadels: Anthropological Interventions in Emerging Sciences and Technologies* (Santa Fe, NM: School of American Research, 1998).

3. Epstein, *Inclusion*.

4. Shapin, *Scientific Life*, 18.

5. Paul Rabinow, *Making PCR: A Story of Biotechnology* (Chicago: University of Chicago Press, 1996).

6. Rabinow, *Anthropos Today*; Shapin, *Scientific Life*.

7. Shapin, *Scientific Life*, 6.

8. *Ibid.*

9. In Duana Fullwiley's study of the Burchard lab at the University of California, San Francisco, Burchard similarly told her: "I think it's our responsibility as physician-scientists, as members of this population, and as taxpayers who are funding the NIH, that we require the NIH to study minority populations." Fullwiley, "Biologistical Construction of Race," 712.

10. Thomas F. Gieryn, *Cultural Boundaries of Science: Credibility on the Line* (Chicago: University of Chicago Press, 1999), 4–5, 15–17.

11. Already 23andMe, with its online community and Relative Finder application, is encouraging consumers to interact with others based on their haplotypic profiles.

12. See Shapin, *Scientific Life*, for a discussion of biotech scientists as late modernity's "New Men and Women."

13. *Economist*, "The Proper Study of Mankind: Practical Applications Are All Well and Good" (July 1, 2000); Edward Rothstein, "Genes, People, and Languages," *New York Times* (Apr. 1, 2000), 9; Mark Ridley, "How Far from the Tree?" *New York Times* (Aug. 20, 2000).

14. Jorde, Bamshad, and other members of the Jorde lab also reported on the proportion of diversity within those clusters. The lab often published reports on population clusters and population diversity in tandem; however, this was the first article to purposefully address ways to get around using proxies for genetic variation.

15. *Alus* are short repeating segments about three hundred nucleotides long. Mobile elements interspersed throughout the genome, they provide an easy way for population scientists to classify humans.

16. Therefore, they argued that thinking in continental terms was useful only for some groups (Europeans, Far East Asians, but not Africans, Central Asians, etc.). Michael J. Bamshad, Stephen Wooding, W. Scott Watkins, Christopher T. Ostler, et al., "Human Population Genetic Structure and Inference of Group Membership," *American Journal of Human Genetics* 72 (2003), 578.

17. Steve Olsen, *Mapping Human History: Genes, Race, and Our Common Origins* (New York: Houghton Mifflin, 2003).

18. Ritter, "As Role of Race in Research Questioned, a DNA Databank Proposed for Blacks."

19. Michael Bamshad, Stephen Wooding, Benjamin A. Salisbury, and J. Claiborne Stephens, "Deconstructing the Relationship Between Genetics and Race," *Nature Reviews Genetics* 5 (2004), 607.

20. An interesting study released at this time was a meta-analysis of how many studies found that gene variants produced different biological effects. This screen of 697 study populations led by Tufts University's John Ioannidis showed that while gene frequencies varied across these study populations at the rate of 58 percent, there was a significant difference in the effect of variants only 14 percent of the time. Ioannidis and his colleagues thus concluded that the biological impact of genes was overwhelmingly consistent across humankind. John P.A. Ioannidis, Evangelina E. Ntzani, and Thomas A. Trikalinos, "'Racial' Differences in Genetic Effects for Complex Diseases," *Nature Genetics* 36 (2004), 1312–18. Though these studies did not make waves in the popular press, they framed much of the debate to come. They also displayed how entrenched was the belief that genomics would settle the question of race for all humankind.

21. Svante Paäbo, "The Mosaic That Is Our Genome," *Nature* 421 (2003), 410.

22. Marcus W. Feldman, Richard C. Lewontin, and Mary-Claire King, "A Genetic Melting-pot," *Nature* 242 (2003), 374.

23. Francesc Calafell, "Classifying Humans," *Nature Genetics* 33 (2003), 424.

24. Rick A. Kittles and Kenneth M. Weiss, "Race, Ancestry, and Genes: Implications for Defining Disease Risk," *Annual Review of Genomics and Human Genetics* 4 (2003), 37.

25. Jeffrey C. Long and Rick A. Kittles, "Human Genetic Diversity and the Nonexistence of Biological Races," *Human Biology* 75 (2003), 449–71.

26. Susanne B. Haga and J. Craig Venter, "FDA Races in Wrong Direction," *Science* 301 (2003), 466.

27. David Rotman, "Genes, Medicine, and the New Race Debate," *Technology Review* (June 1, 2003), 14.

28. Peter Gorner, "DNA/Disease Database on Blacks Raises Fears," *Chicago Tribune* (Feb. 15, 2004).

29. Maggie Fox, "Don't Base Drug Policy on Race, Geneticists Say," Reuters News (July 24, 2003).

30. *Scientific American*, "Racing to Conclusions" (Nov. 10, 2003), http://www.scientific american.com/article.cfm?id=racing-to-conclusions.

31. Mildred K. Cho and Pamela Sankar, "Forensic Genetics and Ethical, Legal, and Social Implications Beyond the Clinic," *Nature Genetics* 36 (2004), S8–S12.

32. Charmaine D.M. Royal and Georgia M. Dunston, "Changing the Paradigm from 'Race' to Human Genome Variation," *Nature Genetics* 36 (2004), S5–S7.

33. Bioethicists Pamela Sankar and Mildred Cho warned about "function creep"—the unintended application of genomic research outside the halls of medicine—using forensic genetics as an example. Sankar and Cho criticized the profusion of consumer genomics, which is seen as determining the race of samples, arguing that innovations can contribute to racist systems. Though Shriver and Kittles had assured that they would do everything in their power to steer customers away from interpreting their genomic data in terms of race (Mark D. Shriver and Rick A. Kittles, "Genetic Ancestry and the Search for Personalized Genetic Histories," *Nature Reviews Genetics* 5 [2004], 616), Cho and Sankar countered that racism in society was so institutionalized that simple warnings would fail to prevent consumers from reading race into their data.

In a later response, Shriver, Frudakis, and Federal Bureau of Investigation officer Bruce Budowle suggested that the population genomics behind their product was sound science and that the precision of DNA data would reduce the need for subjective eyeballing measures like racial profiling. Shriver, Frudakis, and Budowle acknowledged the ethical problems with DNA surveillance in general, but maintained that scientific measures were better tools than eyewitness accounts. Mark Shriver, Tony Frudakis, and Bruce Budowle, "Getting the Science and the Ethics Right in Forensic Genetics," *Nature Genetics* 37 (2005), 449–51. Cho and Sankar replied with further skepticism about the state of the science. They argued that DNA should not be used predictively to create a pool of suspects until individual profiling is foolproof. Mildred K. Cho and Pamela Sankar, "In reply to 'Getting the Science and the Ethics Right in Forensic Genetics,'" *Nature Genetics* 37 (2005), 450–51.

34. Joanna L. Mountain and Neil Risch, "Assessing Genetic Contributions to Phenotypic Differences Among 'Racial' and 'Ethnic' Groups," *Nature Genetics* 36 (2004), S48–S53.

35. E. J. Parra, R. A. Kittles, and M. D. Shriver, "Implications of Correlations Between Skin Color and Genetic Ancestry for Biomedical Research," *Nature Genetics* 36 (2004), S54–S60.

36. Lynn B. Jorde and Stephen P. Wooding, "Genetic variation, classification and 'race.'" *Nature Genetics* 36 (2004), S28-S33; Charles N. Rotimi, "Are Medical and Non-medical Uses of Large-scale Genomic Markers Conflating Genetics and 'Race'?" *Nature Genetics* 36 (2004), S43–S47.

37. S.O.Y. Keita, R. A. Kittles, C.D.M. Royal, G. E. Bonney, et al., "Conceptualizing Human Variation," *Nature Genetics* 36 (2004), S17–S20.

38. Sarah K. Tate and David B. Goldstein, "Will Tomorrow's Medicines Work for Everyone?" *Nature Genetics* 36 (2004), S34–S42.

39. Sarah A. Tishkoff and Kenneth K. Kidd, "Implications of Biogeography of Human Populations for 'Race' and Medicine," *Nature Genetics* 36 (2004), S21–S27.

40. Shriver's graduate student Megan Alicia Rogers also reported this study at the Genetics and Genealogy Symposium (2011).

41. Adam Bostanci, "Genetic Ancestry Testing as Ethnic Profiling," *Science as Culture* 10 (2010), 107–14. Also see Bliss, "Mapping Race Through Admixture."

42. What can only be seen as a reward for creating a racial project that would better align operations with the contemporary racial paradigm, DNAPrint was awarded an NIH Small Business Innovation Research grant to develop the software.

43. "Senecio Software and DNAPrint Genomics Receive NIH Research Grant" (Jan. 31, 2005), http://www.senecio.com/news310105.html. Also see "DNAPrint Teams with Senecio Software to Win NIH-SBIR Grant," http://www.dnaprint.com/welcome/press/press_recent.

44. Sandra Soo-Jin Lee, Joanna Mountain, Barbara Koenig, Russ Altman, et al., "The Ethics of Characterizing Difference: Guiding Principles on Using Racial Categories in Human Genetics," *Genome Biology* 9 (2009), 404.

45. Genomicists also describe meetings as frustrating and contentious. A common reflection suggests that scientists at times feel ill equipped for policymaking. However, faith in genomic taxonomy propels scientists further into policymaking all the while.

46. The Social Issues Committee published a statement in 2007: K. Hudson, G. Javitt, W. Burke, W. P. Byers, with ASHG Committee, "ASHG Statement on Direct-to-Consumer Genetic Testing in the United States," *American Journal of Human Genetics* 81 (2007), 635–37.

47. American Society of Human Genetics, "Ancestry Testing Statement" (Nov. 13, 2008), http://www.ashg.org/pdf/ASHGAncestryTestingStatement_FINAL.pdf; Charmaine D. Royal, John Novembre, Stephanie M. Fullerton, David B. Goldstein, et al., "Inferring Genetic Ancestry: Opportunities, Challenges, and Implications," *American Journal of Human Genetics* 86 (2010), 661–73.

48. *Ibid.*

49. In her research on peer review in the world of research funding, Michele Lamont has noted the recent trend toward characterizing interdisciplinarity in terms of diversity. See Lamont, *How Professors Think.*

50. Bourdieu, *Pascalian Meditations*; *Science of Science and Reflexivity.*

51. All statistics provided by Ronda Britt, "Universities Report $55 Billion in Science and Engineering R&D Spending for FY 2009; Redesigned Survey to Launch in 2010,"

National Science Foundation 10-329 (Sept. 2010), http://www.nsf.gov/statistics/infbrief/nsf10329/nsf10329.pdf.

52. Though counterintuitive to the way we think about the fashioning of coherent identities for emergent sciences, internal discontinuities in practice are also a site of gains for the field. Scientists relying on an every-lab-for-itself taxonomic practice, as described in the previous chapter, are able to continue to investigate and apply race in ways that perpetuate controversy across the field and guarantee the field's autonomy from extrascientific agents like the government. Refusing to apply federal standardizations insulates individual scientists from institutional oversight and the entire field from political oversight. In this way, individual scientists' autonomy is linked to the field's autonomy.

53. Adele E. Clarke, Laura Mamo, Jennifer Fosket, Jennifer Fishman, et al., *Biomedicalization: Technoscience and Transformations of Health and Illness in the U.S.* (Durham, NC: Duke University Press, 2010).

54. Rose, *Politics of Life Itself.*

55. Jenny Reardon suggests that the way users adopt the results of ancestry tests to garner resources—like access to affirmative action college admissions—promotes an individualization of race. Reardon argues that individualizing race takes power away from the very underserved groups for which political racial identities exist. I agree with Reardon's claim, and interpret the simultaneous corruption and enrichment of racial taxonomy as characteristic of the contradictory forces produced by differently situated stakeholders. Reardon, "Finding Oprah's Roots, Losing the World."

56. David B. Goldstein and Huntington F. Willard, "Race and the Genome," *Boston Globe* (Jan. 17, 2005), A11.

57. Nitzan Mekel-Bobrov, Sandra L. Gilbert, Patrick D. Evans, Eric J. Vallender, et al., "Ongoing Adaptive Evolution of *ASPM*, a Brain Size Determinant in Homo sapiens," *Science* 309 (2005), 1720–22; Patrick D. Evans, Sandra L. Gilbert, Nitzan Mekel-Bobrov, Eric J. Vallender, et al., "Microcephalin, a Gene Regulating Brain Size, Continues to Evolve Adaptively in Humans," *Science* 309 (2005), 1717–20.

58. *ASPM* and *Microcephalin* are genes associated with microcephaly (*ASPM* stands for "abnormal spindle-like microcephaly associated"). These studies asserted that positive selection of a haplotype for bigger brain size emerged during the periods of cultural growth outside of Africa:

the age of haplogroup D and its geographic distribution across Eurasia roughly coincide with two important events in the cultural evolution of Eurasia—namely, the emergence and spread of domestication from the Middle East 10,000 years ago and the rapid increase in population associated with the development of cities and written language 5,000 to 6,000 years ago around the Middle East.

59. For a review, see Pardis C. Sabeti, Patrick Varilly, Ben Fry, Jason Lohmueller, et al., and the International HapMap Consortium, "Genome-wide Detection and Characterization of Positive Selection in Human Populations," *Nature* 449 (2007), 913–19.

60. In a subsequent study, a team of neurologists tested variants of the two genes in normal subjects and found no evidence of an association between brain size and

gene variant. The team argued that selection on the genes must have to do with the genes' functions outside of the brain or to subtle neurological effects. Roger P. Woods, Nelson B. Freimer, Joseph A. De Young, Scott C. Fears, et al., "Normal Variants of Microcephalin and ASPM Do Not Account for Brain Size Variability," *Human Molecular Genetics* 15 (2006), 2025–29. In March 2007, Lahn and colleagues recanted their implication that selected-for brain size genes were associated with increases in intelligence. They went so far as to conduct their own test of the association of the brain size genes and IQ, arguing that no correlation existed. Nitzan Mekel-Bobrov, Danielle Posthuma, Sandra L. Gilbert, Penelope Lind, et al., "The Ongoing Adaptive Evolution of ASPM and Microcephalin Is Not Explained by Increased Intelligence," *Human Molecular Genetics* 16 (2007), 600–608. One month later, two laboratories failed to confirm the original studies' findings. Fuli Yu, R. Sean Hill, Stephen F. Schaffner, Pardis C. Sabeti, et al., "Comment on 'Ongoing Adaptive Evolution of ASPM, a Brain Size Determinant in Homo sapiens,'" *Science* 316 (2007), 370b; Nicholas Timpson, Jon Heron, George Davey Smith, and Wolfgang Enard, "Comment on Papers by Evans et al. and Mekel-Bobrov et al. on Evidence for Positive Selection of MCPH1 and ASPM," *Science* 317 (2007), 1036a. In a response article, Lahn and lead coauthor Nitzan Mekel-Brobokov maintained their position on the existence of selection but reemphasized the absence of a correlation with intelligence. Nitzan Mekel-Brobokov and Bruce T. Lahn, "Response to Comments by Timpson et al. and Yu et al.," *Science* 317 (2007), 1036b.

61. Emily Singer, "Beyond Race-Based Medicine," *Technology Review* (Jan. 16, 2009), http://www.technologyreview.com/printer_friendly_article.aspx?id=2197.

62. Karen Bartlett, "Hard to Swallow: Race-Based Medicine," *Wired* (Oct. 20, 2009), http://www.wired.co.uk/magazine/archive/2009/11/features/hard-to-swallow-race -based-medicine.

Chapter 7

1. Sheila Jasanoff, *Reframing Rights: Bioconstitutionalism in the Genetic Age* (Cambridge, MA: MIT Press, 2011).

2. NIH News, "NIH and Wellcome Trust Announce Partnership to Support Population-based Genome Studies in Africa" (June 22, 2010), http://www.nih.gov/news/health/jun2010/nhgri-22.htm.

3. Antonio Regalado, "Indigenous Peruvian Tribe Blocks DNA Sampling by National Geographic," *Science Insider* (May 6, 2011), http://news.sciencemag.org/science insider/2011/05/indigenous-peruvian-tribe-blocks.html. Also see the 2010 debate over research on Havasupai DNA: Amy Harmon, "Where'd You Go with My DNA?" *New York Times* (Apr. 25, 2010), WK1; Jennifer Couzin-Frankel, "DNA Returned to Tribe, Raising Questions About Consent," *Science* 328 (2010), 558.

4. Vanessa M. Hayes, "Indigenous Genomics," *Science* 332 (2011), 639.

5. Rajan, *Biocapital.*

6. Benjamin, "Lab of Their Own."

7. Daar and Singer, "Pharmacogenetics and Geographical Ancestry."

8. Benjamin, "Lab of Their Own; Vivette García, "La renovación del discurso nacionalista y la investigación genómica en México" (unpublished manuscript).

9. GenomeWeb, "NHGRI Awards $18.4M to 10 Groups Developing New Sequencing Technologies," http://www.genomeweb.com/sequencing/nhgri-awards-184m-10-groups-developing-new-sequencing-technologies.

10. See, for example, http://crggh.nih.gov.

11. See *Nature* volume 455; *Nature Genetics* volumes 39–40; and *Science* volumes 319–22.

12. Constance Holden, "The Sociable Brain," *Science* 321 (2008), 487; Gene E. Robinson, Russell D. Fernald, and David F. Clayton, "Genes and Social Behavior," *Science* 322 (2008), 896–900; James H. Fowler and Christopher T. Dawes, "Partisanship, Voting, and the Dopamine D2 Receptor Gene" (2009), http://jhfowler.ucsd.edu; James H. Fowler, "Two Genes Predict Voter Turnout," *Journal of Politics* 70 (2008), 579–94.

13. Barry J. Dickson, "Wired for Sex: The Neurobiology of Drosophila Mating Decisions," *Science* 322 (2008), 904–9; Zoe R. Donaldson and Larry J. Young, "Oxytocin, Vasopressin, and the Neurogenetics of Sociality," *Science* 322 (2008), 900.

14. For reviews, see Holden, "Sociable Brain"; and Robinson, Fernald, and Clayton, "Genes and Social Behavior."

15. Dickson, "Wired for Sex"; Joseph S. Takahashi, Kazuhiro Shimomura, and Vivek Kumar, "Searching for Genes Underlying Behavior: Lessons from Circadian Rhythms," *Science* 322 (2008), 909–12.

16. P. D. Koellinger, M.J.H.M. van der Loos, P.J.F. Groenen, A. R. Thurik, et al., "Genome-wide Association Studies in Economic and Entrepreneurship Research: Promises and Limitations," *Small Business Economics* 35 (2010), 1–18; P. D. Kroellinger, M. van der Loos, P. Groenen, and R. Thurik, "Genome-wide Association Studies and the Genetics of Entrepreneurship," *European Journal of Epidemiology* 25 (2010), 1–3.

17. Jan-Emmanuel De Neve's work on personality and leadership has been talked about in *Psychology Today*, *Mental Health News*, and *The Daily Beast* and presented at a TED talk. Jan-Emmanuel De Neve, "Functional Polymorphism (5-HTTLPR) in the Serotonin Transporter Gene Is Associated with Subjective Well-Being: Evidence from a U.S. Nationally Representative Sample," *Journal of Human Genetics* (forthcoming, 2011); Jan-Emmanuel De Neve, James H. Fowler, Bruno S. Frey, and Nicholas A. Christakis, "Genes, Economics, and Happiness" (unpublished manuscript).

18. Richard Arvey, *Born Entrepreneurs, Born Leaders: How Your Genes Affect Your Work Life* (Oxford, UK: Oxford University Press, 2010).

19. Binghamton University, "Propensity for One-Night Stands, Uncommitted Sex Could Be Genetic, Study Suggests," *Science Daily* (2010), http://www.sciencedaily.com/releases/2010/12/101201095601.htm.

20. Constance Holden, "Do Good Sperm Predict a Good Brain?" *Science* 321 (2008), 487; Ben W. Heineman, "The In-Group Rules," *Science* 319 (2008), 904–5; Peter K. Hatemi, Nathan A. Gillespie, Lindon J. Eaves, Brion S. Maher, et al., "A Genome-Wide Analysis of Liberal and Conservative Political Attitudes," *Journal of Politics* 73 (2011),

271–85; University of California, San Diego, "Researchers Find a 'Liberal Gene,'" *Science Daily* (2010), http://www.sciencedaily.com/releases/2010/10/101027161452.htm.

21. On the shortcomings of molecular frameworks in social and environmental research, see Sara Shostak, *Defining Vulnerabilities* (Berkeley: University of California Press, 2013).

22. Another case concerns the theory on Ashkenazi Jewish intelligence put forth by Henry Harpending, Greg Cochran, and others. See Gregory Cochran, Jason Hardy, and Henry Harpending, "Natural History of Ashkenazi Intelligence," *Journal of Biosocial Science* 38 (2006), 659–93.

23. Florida State University, "'Warrior Gene' Linked to Gang Membership, Weapon Use," *Science Daily* (2011), http://www.sciencedaily.com/releases/2009/06/090605123237.htm; John Saunders, "Genetic Link to Credit Card Debt" (May 4, 2010), http://www.independent.co.uk/news/science/genetic-link-to-credit-card-debt-1961483.html

24. Brown University, "'Warrior Gene' Predicts Aggressive Behavior After Provocation," *Science Daily* (2009), http://www.sciencedaily.com /releases/2009/01/090121093343.htm.

25. John Horgam, "Have Researchers Really Discovered Any Genes for Behavior? Candidates Welcome," *Scientific American* weblog, Cross-check (May 2, 2011), http://www.scientificamerican.com/blog/post.cfm?id=have-researchers-really-discovered-2011-05-02.

26. Rose McDermotta, Dustin Tinglyeb, Jonathan Cowdenc, Giovanni Frazzettod, et al., "Monoamine Oxidase A Gene (*MAOA*) Predicts Behavioral Aggression Following Provocation," *PNAS* 106 (2009), 2118–23. McDermotta et al. cited Y. Gilad, S. Rosenberg, M. Przeworski, D. Lancet, and K. Skorecki, "Evidence for Positive Selection and Population Structure at the Human *MAO-A* Gene," *PNAS* 99 (2002), 862–67.

27. Rod Lea and Geoffrey Chambers, "Monoamine Oxidase, Addiction, and the 'Warrior' Gene Hypothesis," *Journal of the New Zealand Medical Association* 120 (2007), http://www.nzmj.com/journal/120-1250/2441.

28. The Council of Conservative Citizens republished these statistics on their website as:

American black males are twice as likely than American white males to have *MAOA-L* which has been linked to crime, violence and aggression in scores of studies going back over fifteen years. Black males are also 13.5 times more likely to have a rare version of the gene associated with "extreme violence and extreme aggression." Latinos and American Indians are also nearly twice as likely as whites to have the more common version of the gene. However they are only about one fourth as likely to have the extreme version compared to blacks.

See http://cofcc.org/2010/05/science-daily-whites-have-lowest-instance-of-maoa-l-gene-which-is-linked-to-aggression-violence-crime-and-sexual-abuse.

29. One 2011 *Science Daily* article announced "deep" differences between whites and Asians. On closer inspection, the article reported on a merely routine psychology finding that patterns in cultural perceptions are reflected in brain activity. Association for Psychological Science, "Cultural Differences Are Evident Deep in the Brain

of Caucasian and Asian People," *Science Daily* (2011), http://www.sciencedaily.com/releases/2011/04/110411163922.htm. Also see University of Colorado, Denver, "Could Brain Abnormalities Cause Antisocial Behavior and Drug Abuse in Boys?" *Science Daily* (2010), http://www.sciencedaily.com/releases/2010/09/100923104212.htm.

30. Jon Entine, "The Straw Man of Race," *World and I* 16 (2001), 294–317.

31. Weekly Bulletin, "Academic Study Attempts to Explain Why Black Athletes Are Faster," *Journal of Blacks in Higher Education* (Aug. 5, 2010).

32. *Ibid.*

33. NCAA data for the 2008–09 academic year predicted that the policy would affect more than 166,900 athletes. Katie Thomas and Brett Zarda, "In N.C.A.A., Question of Bias over a Test for a Genetic Trait," *New York Times* (Apr. 11, 2010), D1. Also see William Saletan, "Lose the Race: Can a Black-White Performance Gap Be Hereditary but Not Racial?" *Slate* (July 13, 2010), http://www.slate.com/id/2260314.

34. Troy Duster, "Selective Arrests, an Ever-Expanding DNA Forensic Database, and the Specter of an Early-Twenty-First-Century Equivalent of Phrenology," in *Technology of Justice: DNA and the Criminal Justice System*, edited by David Lazer (Cambridge, MA: MIT Press, 2004).

35. Christian Torres, "Telemedicine Has More than a Remote Chance in Prisons," *Nature Medicine* 16 (2010), 496.

36. Take, for instance, 23andMe's research into Parkinson's. Most personal genomics companies sell (or primarily sell) their products only in the United States, Europe, and Canada. Elizabeth Pennisi, "Number of Sequenced Human Genomes Doubles," *Science* 322 (2008), 838.

37. Atul J. Butte, "The Ultimate Model Organism," *Science* 320 (2008), 325–27; Richard G.H. Cotton, Arleen D. Auerbach, Myles Axton, Carol Isaacson Barash, et al., "The Human Variome Project," *Science* 322 (2008), 861–62; Elizabeth Pennisi, "Proposal to 'Wikify' GenBank Meets Stiff Resistance," *Science* 319 (2008), 1598–99.

38. Benjamin Weiser, "Court Rejects Judge's Assertion of a Child Pornography Gene," *New York Times* (Jan. 28, 2011), A20.

39. Barbara Bradley Hagerty, "Can Your Genes Make You Murder?" NPR (July 1, 2010), http://www.npr.org/templates/story/story.php?storyId=128043329.

40. See, for example, Laura Bevilacqua, Stéphane Doly, Jaako Kaprio, Qiaoping Yuan, et al., "Did HTR2B Make Them Do It?" *Nature* 468 (2010), 1061–66.

41. John Travis, "Scientists Decry 'Flawed' and 'Horrifying' Nationality Tests," *Science Insider* (2009), http://news.sciencemag.org/scienceinsider/2009/09/border-agencys.html.

42. Katie Drummond, "Genetic Patdown," *The Daily* (Feb. 26, 2011), http://www.thedaily.com/page/2011/02/26/news-tsa-scanner-1-2.

43. The burgeoning research programs of proteomics and epigenomics, which study genetic process regulation and expression shaped by the environment, have thus far envisioned the environment in terms of chemical processes; thus, though they have the potential to introduce a social perspective on behavior, they have remained within the logic of essential biological difference. Governmental and lay racial categories abound in the studies put forth by these emerging fields.

44. Matthew Hughey, *White Bound: White Nationalists, White Antiracists, and the Shared Meanings of Race* (Stanford, CA: Stanford University Press, 2012). Also see Eduardo Bonilla-Silva, *Racism Without Racists: Color-blind Racism and the Persistence of Racial Inequality in the United States* (Lanham, MD: Rowman and Littlefield, 2006).

45. See, for example, Rajesh Kumar, Max A. Seibold, Melinda C. Aldrich, L. Keoki Williams, et al., "Genetic Ancestry in Lung-Function Predictions," *New England Journal of Medicine* 1056 (2011), 1–10; and Charles N. Rotimi and Lynn B. Jorde, "Ancestry and Disease in the Age of Genomic Medicine," *New England Journal of Medicine* 363 (2010), 1551–58. On the prevalence and quality of ancestry as a replacement, see Fujimura and Rajagopalan, "Different Differences."

46. See Angelo Falcón, "Latinos, Diversity, and Racial Fatigue in the Age of Obama," *National Civic Review* 98 (2009), 40–42.

47. Lundy Braun and Evelynn Hammonds describe the way population alternatives to race, such as the sub-Saharan African tribal taxonomy proffered for the Human Genome Diversity Project, have political genealogies that stretch across centuries. Lundy Braun and Evelyn Hammonds, "Race, Populations, and Genomics: Africa as Laboratory," *Social Science and Medicine* 67 (2008), 1580–88.

48. George T.H. Ellison, Andrew Smart, Richard Tutton, Simon M. Outram, et al., "Racial Categories in Medicine: A Failure of Evidence-Based Practice?" *PLoS Medicine* 4 (2007), 134–36.

49. Kenan Malik, "The Science of Race and the Politics of Ignorance," *Philosophers' Magazine* 41 (2008).

50. This is a point Jenny Reardon hoped for in her concluding remarks about the failure of the Human Genome Diversity Project. Reardon, *Race to the Finish.*

51. Sarah Richardson, "Race and IQ in the Postgenomic Age: The Microcephaly Case," *Biosocieties* 6 (2011): 420–46.

Index